The Development of High Technology Industries

An International Survey

Edited by
MICHAEL J. BREHENY and RONALD McQUAID

CROOM HELM
London • New York • Sydney

© 1987 Michael J. Breheny and Ronald McQuaid
Croom Helm Ltd, Provident House, Burrell Row,
Beckenham, Kent, BR3 1AT
Croom Helm Australia, 44-50 Waterloo Road,
North Ryde, 2113, New South Wales

British Library Cataloguing in Publication Data
The Development of high technology industries:
   an international survey.
   1. High technology industries.
   I. Breheny, M.J. II. McQuaid, Ronald
   338.4'76   HC79.H5
   ISBN 0-7099-3942-6 ✓

Published in the USA by
Croom Helm
in association with Methuen, Inc.
29 West 35th Street
New York, NY 10001

Library of Congress Cataloging-in-Publication Data
The Development of high technology industries.

   Includes index.
   Contents: High technology industry and the international
division of labour / A. Sayer and K. Morgan —
The growth and internationalisation of the American
semiconductor industry / J. Henderson and A.J. Scott —
Technology waves and the future of employment and wealth
creation in Britain / P. Preston — [etc.]
   1. High technology industries. I. Breheny, M.J.
(Michael J.) II. McQuaid, Ronald, 1955–   .
HC79.H53D48 1987   338.4'762   87-9086
ISBN 0-7099-3942-6

Printed and bound in Great Britain
by Billing & Sons Limited, Worcester.

CONTENTS

Chapter One

INTRODUCTION

Michael Breheny and Ronald McQuaid

There is little doubt that in just about all western countries the industrial phenomenon that has caused most excitement in the late 1970s and 1980s has been so-called 'high technology' industry. The phrase, or more usually its colloquial form 'high tech', has become synonymous with all things modern and progressive; not just industrial products and processes, but all manner of artefacts and activities; from buildings, to methods of childbirth, to running shoes.

Whilst some of this labelling may be trivial, the concern in industry with high technology has been deadly serious. Some countries, notably Japan and the United States appear to have based major economic revivals on their prowess in developing and using high technology products. Some newly industrialising countries, particularly in South East Asia, have become major manufacturers of these products, albeit often for foreign companies. Other, western countries, such as Britain, have looked upon high technology sectors as the potential saviours of their depressed economies; 'either this or a return to a peasant economy', is the message often heard. This prospect of promoting high technology industries to revitalise flagging economies has been seized upon also by local governments in many countries, who focus on this sector in their local industrial promotion initiatives.

Thus, high technology is a serious business. However, despite all of this concern and activity, and this use of the common all-purpose label, there is no guarantee that commentators from different countries are talking about the same phenomenon, or that they all attach the same importance to the issue. Nor have we a grasp of respective experiences in high technology industries - however defined - at

1

national, regional and local levels.

Given the apparently crucial role that high technology industry does or ought to play in many countries, this failure to make use of different experiences is alarming. It is alarming for both academic and practical reasons. Our understanding of the growth and development of high technology industries, from global to local scales, can only benefit from knowledge of different experiences. Likewise, practical responses will be more effective with such knowledge. If some countries or localities have discovered ways of promoting genuine growth in these sectors, then others may be able to learn from this experience and avoid the anguish of experimentation. If there is evidence that high technology industry provides remarkably few jobs, then this might influence the attitudes and policies of others. If more jobs appear to be produced in particular circumstances or from particular high tech sectors, then again the information is valuable. If there is evidence that the growth of high technology industries has particular labour market effects - benefiting some groups at the expense of others - then this may well affect the promotional strategies adopted by others. Knowledge of the spatial, global strategies adopted by high technology companies might well influence both theoretical understanding and practical responses. In turn, greater theoretical understanding and practical knowledge feed from each other.

This book makes one modest contribution to this pooling of knowledge. It attempts to bring together, from a set of papers specially commissioned for the purpose, views on the growth and development of high technology industry from a number of different national and international perspectives. The papers were deliberately commissioned in two groups; one consisting of international perspectives on high technology industry, the other of national experiences. The first group helps to set a context for the second because, as we shall see, national roles can only really be understood within their international setting. Thus, we have contributions focusing on the internationalisation of the electronics and telecommunications industries, within an 'international divisions of labour' perspective, and national contributions, picking up many of the same issues, from Japan, Canada, France, Australia, and the United Kingdom.

Clearly, these different national perspectives provide only a sample - and one biased to advanced industrial nations - of those that we might consider important to our consideration of high technology industry. However, they do at least go some way to rectifying the shortage of cross-national reviews that we have identified. The bias in our sample of national perspectives is to some degree overcome by the international assessments, which bring the role of a number of other countries, particularly newly-industrialising ones, into the picture.

The first two of our international perspectives, by Sayer and Morgan and Henderson and Scott, both consider the internationalisation of high technology industries - electronics generally and semiconductors, respectively - from their original, core areas. Sayer and Morgan address the international division of labour in the electronics field, covering three independent but converging sectors - semiconductors, computer systems and telecommunications. Discussion of these sectors themselves is preceded by a presentation of some crucial theoretical points; on the role of technology in competition, and on the international division of labour. On the latter issue they stress two points. They criticise the excessive reliance of much radical theory on labour as the only 'locational factor' explaining corporate spatial strategies. They also criticise the failure of such theory to consider the role of nation states; too often researchers assume a 'stateless, laissez-faire world economy in which nation-states simply formed a passive backcloth'. They also warn against the construction of neat, formal notions of spatial divisions of labour (cf. Henderson and Scott), arguing that the international division of labour 'has emerged historically and largely unintentionally out of a myriad of decisions made by governments and firms'. Assessments of the internationalisation of high technology industry from its cores in the U.S.A., Europe and Japan follow, along with specific examples of corporate spatial divisions of labour. The paper - intended as an 'antidote to the parochialism' of much current work - concludes with a number of specific critical comments on research.

Henderson and Scott also place the growth of high technology industry - in this instance the American semiconductor industry, a 'paradigmatic case' - within a broader consideration of changing spatial divisions of labour. They trace the 'historical and geographical trajectory' of the industry from its roots in California to its use of peripheral locations around the world for the

3

carrying out of specific labour processes, especially assembly and testing. The analysis of this internationalisation is pursued through case studies of two locations where the phenomenon is particularly marked; South East Asia and Scotland. Henderson and Scott stress the importance of distinguishing between 'internal necessary relations', the basic, fundamental nature of capitalist production, and 'external contingent relations', which may help to determine its particular form and location at any point in time. All too often, they claim, researchers emphasise the latter to the exclusion of the former in their attempts at explanation. They conclude with the message that familiar spatial divisions of labour - of the kind they describe in their case studies - are themselves only transitory features of capital's ever changing search for advantage on a global scale.

Gillespie, Howells, Williams and Thwaites pick up the internationalisation theme again: in this case focusing on the Information Technology (IT) industries in the 'less favoured regions of Europe'. Here, however, corporate responses to increasing competition are emphasised, with an attempt to rectify the usual, but excessive stress on price competition by explaining the crucial role of product innovation. They explain the imperatives of innovation in IT industries and the inevitable inflationary effects these have on R and D expenditure, on viable levels of operation, on process automation and on the necessary globalisation of markets.

The problems faced by European companies and countries, in an industry in which massive investments and massive markets are required, are debated. The response, in the form of the ESPRIT programme and joint corporate ventures particularly, is explained. The battle to stay in the global IT 'top league' with the Americans and Japanese will depend on the success of European consortia. The corporate organisational responses to intensified competition - for example the move to JIT (just-in-time) manufacture - are discussed, along with locational and labour market changes. The paper concludes with suggestions for elements of an IT production strategy for 'Europe's less favoured regions'; the attraction of transnational corporation production facilities, the building up of 'national champion' companies, and the identification and exploitation of 'niche' markets.

Just as Gillespie et al conclude on a

prescriptive note, so Preston begins with questions about the consequences of new technologies for future employment and economic structures. Ultimately, his focus is on such consequences for Britain, but much of his analysis applies equally well in other countries. As such, the paper neatly bridges the international and national perspectives of other contributions. In attempting to find an adequate framework for understanding the nature and consequences of technological change, Preston assesses in some detail the popular conceptions of Stonier and Toffler. Having appraised critically these two views, he suggests that a 'long waves' approach 'offers a more coherent, realistic and less sensationalist perspective'. Modern versions of long waves theory, which reject 'technological determinism' and stress the 'dynamic harmony' between technological change and social and economic relations are explained. This approach, it is argued, gives a sounder basis upon which to speculate about future change, which is likely to focus on the convergence of new technological advances in microelectronics, computing and telecommunications. The implications of such changes for occupational structures, and for the possibly crucial role of the information service sectors, are discussed.

In the international scramble for position in the high technology stakes, one of the apparent losers amongst western industrialised countries has been Canada. Britton's contribution assesses the magnitude and reasons for the country's 'technology gap', as he terms it. He addresses three major issues in his paper.

First, the historical background to Canada's relatively poor performance in the high technology sectors is assessed. A crucial factor in explaining this performance has been Canada's vulnerability to foreign direct investment - 'industrialisation by invitation' - and the consequent lack of indigenous investment. This analysis is supported by a detailed assessment of Canadian R & D expenditure, of manufacturing performance in defined high technology sectors, of the relative health of Small to Medium Establishments (SMEs) and Large establishments, and of Canada's international trading position. These assessments are followed by a review of the structural problems of Canadian industry and of the significance of the 'technology gap'.

Britton's second major theme takes the analysis down a spatial scale to look at the geography of Canadian technological activity. The dominance of the Toronto region, rather than the more popularly accepted Ottawa concentration, is established. The paper finally focuses on the third theme; that of the incompatibility of the various trends identified in the first two themes with federal and Ontario government industrial programmes. A detailed critique of recent policies is given, and shortcomings identified. For example policies generally fail to target both the right types of firms - no distinction is made between foreign and domestically-controlled firms, for instance - and the locations with most potential, such as Toronto.

The contribution by Pottier sets out to look specifically at the location of high technology industry in France, and to attempt some explanation of the pattern that has emerged in recent years. A dual approach is adopted, in which aggregate and sectoral assessments are made. A definition of high technology sectors, based on skill levels and R&D expenditure, is adopted. The differing geographies of the four sectors studied - microelectronics telecommunications, robotics and aeronautics - and the different forces affecting change in each case, suggest that individual sectors need separate attention. Particular emphasis is placed on the forces which have centralising and decentralising effects. For example, it is demonstrated that requirements for skilled workers tend to have a centralising effect. Paris, for instance, remains a major focus of high technology activity, with less decentralisation of high technology employment than for industry as a whole. Some sectors, such as telecommunications, have seen considerable decen-tralisation from traditional industrial areas; remarkably, Brittany, with some state intervention, has benefited considerably in this sector. Other sectors, in particular aeronautics, have remained heavily concentrated in the Paris region. The paper concludes with a review of the major factors that appear to have influenced locational change.

Rather than location, Macdonald takes policy issues as the main theme of his review of high technology industry in Australia. He explains that the whole notion of high technology activity in Australia is closely bound up with government policy; to the point that the volume of policy is often taken as a measure of the activity itself! The problem of policy is exacerbated, as elsewhere of course, by confusion over different assumptions about the meaning of the term 'high technology'. The supposed benefits of high technology are discussed critically, even sceptically, and an attempt made to put these - employment creation, wealth creation, economic restructuring, environmental and health improvements, etc - into perspective. The potential costs, particularly those that, aided by policy, might distort the economy generally, are assessed. The enthusiasm for fundamental policy shifts to favour high technology industry are contrasted with the inappropriateness and weak intellectual basis of much actual policy at federal and state levels. A detailed review of industrial policies at this latter level is presented, followed by a concluding review of Australia's long term high tech prospects.

The paper by Nishioka and Takeuchi provides a rare insight for English-reading researchers into the internal structure and location of Japanese high technology industry. The paper begins with an identification of those industries regarded as high technology and a general review of post-war, and particularly post 1973 oil crisis, changes in these and related sectors. The changing geographical distribution of these industries is described. These geographical changes are related to changing corporate structures and organisation and to changing products. The emerging spatial distribution of functions bears a strong resemblance to those national spatial divisions of labour discussed in other papers. Whilst considerable decentralisation of manufacturing has taken place to many parts of Japan, the major metropolitan areas, and particularly Tokyo, retain much of the higher order R and D and management functions. Even important new centres of high technology, such as Kyushu - 'Silicon Island' - are predominantly manufacturing centres. The paper discusses in some detail the role of the Tokyo region in all of these changes, and in particular describes the functional specialisation of certain central and suburban areas. A novel feature of the paper, and indeed of Japan, is the explanation of the role of government designated high technology communities - the 'Technopolis' concept. Areas such as Nagaoka and Tsukuba Science City seem to have taken on this role successfully. It is an interesting thought for Japan's free enterprise competitors that not only does the Japanese government plan the nation's high technology industry sectorally but spatially as well.

The paper by Breheny and McQuaid aims to assess the factors - 'contingent' factors under Henderson and Scott's distinction - which have led to the locational concentration of much of the United

7

Kingdom's high technology industry, and particularly
the electronics sector, in the 'Western Crescent'
area to the west of London. The paper begins,
however, with a review of national and regional
performance in defined high technology industries.
The national performance in these crucial sectors is
shown to have been very poor, certainly compared to
that in the United States. Moving down a spatial
scale, the regional geography of aggregate and
individual high tech industries is examined. This
helps to identify a 'Western Crescent' with a more
intense concentration in the eastern end of the M4
motorway, around the county of Berkshire. The role
of Berkshire in the national spatial division of high
technology labour is examined briefly. Two major
findings of the examination of the growth of high
technology industry in the Crescent are the deep
historical roots of the phenomenon, stretching back
to the 1930s, and the crucial role played - largely
accidentally it must be said - by public sector
initiatives. The lessons of the work for policy
formulation, including the possibility of a spatial
component to high technology policy (cf. the Japanese
case), are examined.

It would be very difficult to draw any profound
conclusions from the collection of international and
national reviews of high technology industry
presented here; and hence no attempt will be made.
However, some simple, yet valuable points are clear.
There is a wealth of experience available that should
be tapped to the full. Theoretical and practical
approaches adopted by different researchers and
policy makers do differ fundamentally. For example,
very different degrees of emphasis are placed on
external, determining and internal, contingent
explanations of the growth and location of high
technology industries. These different approaches
may reflect different intentions on the part of the
respective researchers. Those taking the former
stance are more likely to be interested in the
development of theory, whilst those adopting the
latter approach may ultimately be interested in local
policy formulation. The nature of the debate in these
papers, however, suggests that it would be unwise for
anyone to rely solely on one of these approaches. For
example, Sayer and Morgan, whilst clearly attempting
to develop sound theoretical work within a spatial
division of labour perspective, do stress the
necessary, and sometimes overwhelming, role of
contingent factors. Macdonald on the other hand, with
his main interest in policy analysis and formulation,

warns against the myopia, and inevitably poor practical responses, that can arise from parochial and atheoretical considerations of high technology industry. Hopefully, the papers collected here will aid the appreciation of the variety of experiences and views on high technology industry available and force broader, and ultimately more rewarding, research perspectives.

Chapter **Two**

HIGH TECHNOLOGY INDUSTRY AND THE INTERNATIONAL DIVISION OF LABOUR: THE CASE OF ELECTRONICS

Andrew Sayer and Kevin Morgan, School of Social Sciences, University of Sussex

Introduction

Electronics is one of the most internationalised of major industries and its global division of labour is of interest both in itself and as an antidote to the parochialism which often accompanies the debate on 'high-tech' in certain regions and localities. In this chapter we shall try to outline and explain this international division of labour, illustrating the arguments by reference to examples drawn from three of the most clearly 'high technology' subsectors, namely, semiconductors, computer systems and telecommunications. These subsectors are not randomly selected, for together they constitute the core of any reasonable definition of 'information technology'; there is also a marked convergence between them. Nevertheless, as we shall see, some significant differences among them remain, in particular the contrast between the free and highly internationalised markets for standard semiconductors and the more nationally-based closed and national markets for many telecommunications products.

We will start by presenting some preliminary theoretical points concerning the role of technology in competition and the study of divisions of labour at the international scale. This will be followed by an account of the rise of the main geographical core areas of strength in the industry and the development of an international division of labour out of these. We will then give some examples of corporate spatial divisions of labour and briefly comment on Britain's place in the international division of labour in electronics. The paper concludes with a review of some of the main theoretical implications of our account.

Our first theoretical point concerns a common

tendency to suppose that high technology industry is
exempt from economic forces - particularly recession
- which affect other industries. Where markets and
new firms are expanding extremely rapidly it is easy
to forget that this is largely an infant industry
phenomenon, and as such vulnerable to a 'shakeout' of
firms unable to sustain a series of innovations and
to produce and market them in rapidly growing
volumes. While it is true that competition in
electronics is exceptionally dependent on product
innovation, recent events have made it abundantly
clear that highly advanced technological innovations
are neither sufficient nor sometimes even necessary
to guarantee success. The astonishing displacement
of a host of small, highly innovative microcomputer
firms by the rise of IBM's technologically-
pedestrian Personal Computer shows that the ability
to produce and market cheaply, and reliably, at high
volume and to provide the security of long-term
customer support counts for more than technological
wizardry.
   It has also been frequently implied that
recession is something that only affects older, low
technology 'sunset' industries, and not new
'sunrise', 'high-tech' industries. Again recent
events show this to be gravely mistaken. Despite
growth rates of over 30% per year in some products,
the semiconductor industry has been extraordinarily
vulnerable to violent fluctuations between booms and
slumps, with frenetic attempts to add new capacity in
the booms leading to massive overproduction and a
precipitous slide into recessions involving
redundancies and closures. Similarly, at the time of
writing (mid-1985), there is a less spectacular but
serious 'shakeout' in the computer system industry.
More generally, such cases remind us that the overriding driving and regulating force is not technology
but capital accumulation. Rapid growth does not
necessarily protect all participants from failure;
on the contrary, it increases the likelihood of
instability. It is about time that these points were
properly appreciated by policy-makers.
   Our second preliminary point or set of points
concerns ways of understanding the 'international
division of labour' (IDL). Interest in the IDL
actually pre-dated work at the sub-national scale on
spatial divisions of labour (Massey, 1984). One of
the most stimulating sources of ideas on the IDL was
the late Stephen Hymer, who wrote in 1975:

   ... a regime of North Atlantic Multinational

Corporations would tend to centralise high level decision-making occupations in a few key cities in the advanced countries, surrounded by a number of regional subcapitals, and confine the rest of the world to the lower levels of activity and income, i.e. to the status of towns and villages in a new Imperial system. Income, status, authority and consumption patterns would radiate out from these centres along a declining curve, and the existing pattern of inequality would be perpetuated. The pattern would be complex, just as the structure of the corporation is complex, but the basic relation between different countries would be one of the superior subordinate, head office and branch plant (Hymer, 1975, page 38).

The parallels with recent work at the subnational scale are clear. Other theories of the IDL of note are Vernon's product life cycle theory and Frobel et al's New International Division of Labour thesis (Vernon, 1966; Frobel et al, 1980). We shall review some of the main problems with these approaches in the conclusion. Suffice it to say for now that they all suffer from a kind of 'tunnel vision' in which a small subset of the whole range of possible forms are treated as universal. As we shall see, actual patterns are far more diverse, and the forces behind them more complex than these theories allow.

For those familiar only with the literature at a subnational scale it is important to note that qualitative differences in divisions of labour and their determinants are to be expected at the international scale. Most obviously, we can expect much greater variations in operating conditions facing firms, particularly labour costs, between countries than within them. Less obviously, there are greater differences in the social organisation of production at the larger scale and these have significant effects on competitiveness and this is so particularly in the case of Japan. However, the main point we want to stress concerns the role of nation states. Too often, the internationalisation of capital is discussed as if it occurred in a stateless, laissez-faire world economy in which nation-states simply formed a passive backcloth (e.g. see Murray, 1972 and Radice, 1975). This is far from the truth, particularly where 'high technology' industry is concerned. The character of national economies, especially that of the home country of a multinational company (MNC) heavily constrains and enables what that firm can do. Also, state policies towards industry (both domestic and foreign), actively shape the form of the international division of labour. These policies range from nationalisation, state procurement, R&D and educational policy, imposition of technical standards, to customs rules and policy on imports and inward investment. Despite the fact that multinationals of the same domicile may be competitors, and hence 'hostile brothers', they have certain interests in common deriving from their shared home base and position with respect to foreign competition. It has been a common fault of analyses of the internationalisation of capital to mistake what are in fact nationally-imprinted characteristics (usually of U.S. MNC's), for the general characteristics of capital pure and simple, in its most advanced form. The implication of this then is that firms are not the only actors in the play: nation states also play a significant part.

It must also be kept in mind that the concept of an international division of labour is inherently ambiguous: it immediately makes one think of a global division of labour wholly structured by multinational firms, but even without any multinational firms, it would be possible to speak of an IDL arising from the different patterns of activity within particular countries. What must be remembered is that the first type of IDL has not replaced the second, but has overlain it. In this paper we shall be concerned with both, on the grounds that even if one is only interested in MNCs, other firms must be taken into account because they have interacted with the MNCs and influenced their behaviour and because the character of national economies and policies has also been influential.

Finally, we must also remember that the IDL has emerged historically and largely unintentionally out of a myriad of decisions made by governments and firms. It is all too easy to misinterpret sequences of limited and piecemeal actions as consequences of conscious global 'strategies'; even where they do exist, they are always subject to continual revision and displacement by contingencies such as unexpected opportunities to buy up firms in trouble, unforeseen downturns in demand or new political regimes.

Having clarified these points, let us now turn to a historical outline of the emergence of the international division of labour in electronics.

The Development of the Industry in the Core Countries
The international division of labour in electronics
has developed from three main bases - USA, Europe and
Japan. As it is vital to appreciate their
characteristics in order to understand the IDL, we
shall begin by describing each in turn. Table 2.1
gives some indication for the relative size of these
three bases and the trade between them for
semiconductors.

Table 2.1: Worldwide Production and Trade in
Integrated Circuits, 1982

	Product-ion	Internal Consumption	Exports	Imports
U.S.	9.3	7.52	1.78	0.58
Japan	3.13	2.23	0.90	0.36
W. Europe	0.79	0.65	0.14	1.25
Rest of the World	0.16	0.13	0.03	0.66

Source: Integrated Circuit Engineering Corp. in
Financial Times 19.8.1983.

Note:
(1) the low consumption and domestic production of
     W. Europe and the failure of the runaway
     industries to make much impact on Rest of the
     World exports
(2) much of the trade will be internal to MNCs.

USA The most striking aspect of the American context
is the enormous size and sophistication of the
national market, both consumer and industrial, but
this did not arise simply by the free play of market
forces. Contrary to the 'strong market-weak state'
stereotype, the U.S. Federal government, via the
Department of Defense and NASA, has historically
played a critical role in the development of the
country's electronics industry. For example the
Department of Defense sustained early supply (by
funding R&D as well as production capacity) and
demand (by being the principal purchaser) (Nelson,
1982). In 1962 the U.S. government accounted for 100%
of integrated circuit sales in the U.S. (McKinsey,
1983). In short, the government provided a vital base
load for the firms to augment their production
capacity and, later, to diversify into commercial
markets. Where government defence procurement in
European countries created only small niches which

14

didn't induce volume production, the scale and duration of the US projects were sufficient to encourage volume production capability as well as technological virtuosity. Another state-related factor has been the high degree of interaction between firms and the scientific establishment afforded by such favourable institutional contexts as the Bell Laboratories. Not surprisingly, the vast majority of key innovations in electronic components and computers have come from the States (Soete and Dosi, 1983). In addition to this, the enormous size of the domestic market allowed synergies to develop between related subsectors. For example, the huge telecommunications and computer markets created the demand for a progressive and efficient volume components sector and in turn this provided the building blocks for the larger 'systems products'.

Europe: Perhaps the most striking contextual contrast with the USA is the fragmentation of the European market into small national markets, each with its own technical standards, its own policies towards imports and inward investors (despite the EEC) and its own government procurement policies, etc. A lower degree of sophistication is also apparent in the market, particularly for computer systems. This fragmentation has inhibited the process of industrial concentration and consequently few European firms approach the US ones in size. Perhaps the most extreme example, but no less important for that, is the difference between the largest European computer producer, Siemens, with a computer revenue of $2.8 billion in 1984, and IBM, with $44.3 billion revenue. This has made it harder to achieve economies of scale and proficiency in volume production while the greater number of protected markets has reduced the spur of competition.

The effects of government protection are clear in the case of the national telecommunications markets. Although the telecoms sector is now under intense political and corporate pressure to 'open up' - a process which is most advanced in North America and the UK - less than 20% of the public network equipment sector worldwide is open to international competitive supply. Procurement and the setting of technical standards within each country is still dominated by PTTs - the national telecommunications authorities - and decision-making is highly politicised.

Although this situation is general outside

Europe, the small size of the European countries has left it with a mosaic of different technical systems and closed national markets, though with the liberalisation of telecommunications in Britain and similar (though more limited) moves elsewhere, the barriers between markets are lowering. This protection has made it difficult for the major national telecommunications firms to penetrate foreign markets, particularly where, as in Britain, the PTT over-specified products, thereby making them difficult to sell abroad. It is significant that one of the leading European telecommunications firms in exporting, Ericsson of Sweden, was exceptional in that it was given specifications by the Swedish PTT which took into account adaptability for other markets. Its electronic exchanges have now far outsold Britain's 'System X' abroad.

A related effect of the small size of national markets and the weakness in volume production is the relative absence of synergies between subsectors. Limited computer and telecommunications markets have created relatively small markets for components firms. Many of the main European electronics firms developed as vertically-integrated electrical companies and newer products such as integrated circuits were able to find in-house uses instead of seeking out more competitive markets. Targeting custom rather than standard products put the firms in a reactive role of responding to customers' individual needs instead of developing a long term strategy to make compatible products (Sciberras et al, 1978). The limited volume production capability of the component firms and frequently also the backwardness of their products have obliged European systems companies to become increasingly reliant on American and Japanese products, whether bought in or made under license. Of course, the small size of the European markets also represents a problem for US and Japanese exporters and inward investors, but in their case they have had the advantages of a larger home market base in which to develop and from which to launch export and foreign direct investment drives.

In Europe also, while defence firms have been quite successful in exporting, the gulf between defence production and civil mass markets is larger than in the U.S. Defence work is generally a small batch or project production, putting specification before price and production process efficiency, and operating within largely protected markets. Civil markets are generally larger and more standardised and are based on keen price competition through volume production. This institutional and cultural gulf (Maddock, 1982) has proved to be virtually unbridgeable for firms accustomed to working within small but comfortable defence market niches; as we shall see it has always been easier to cede the large mass markets to foreign producers than to contest them.

Another important characteristic of the European situation is its early penetration by foreign, especially American, direct investment in electrical engineering and electronics (e.g. Westinghouse, ITT, IBM, Texas Instruments, Burroughs, Sperry). This inward investment was not simply a response to markets but to the tariff barriers which protected them from imports. In the pre-electronic period, in common with the major European firms, many of these formed a stable electro-mechanical oligopoly which lasted for many decades. The American firms led in technology but the European companies were able to maintain their position by imitating the former's products. The break up of the oligopoly began with the introduction by the American firms of integrated circuits (i.c.'s) which suddenly widened the technology gap and hence the imitation lag in Europe. The U.S. firms then began to produce and sell new products in Europe, in some cases establishing new pan-European technical standards and thereby cutting out their European imitators in the market. The situation for the European firms deteriorated because many i.c.'s replaced rather than supplemented the discrete components in which they had had considerable strength. The U.S. companies also started a price war in Europe (though not in America), which further advanced their position. Philips and Siemens were the only European firms large enough to approach the American firms' R&D and investment expenditures and therefore to retain a competitive toe-hold. And even though most other European firms withdrew into niche markets, the American inward investors were often able to penetrate these too.

Japan: Here the electronics industry received no stimulus from the defence sector and at first it lacked a rich technically-sophisticated domestic market. However, by a vigorous process of imitation and improvement behind national barriers to inward investment and licensing (and until 1974, to imports of semiconductors), it became a major force in the world market.

It also benefited from the developments in the social organisation of production in leading firms in the 1960s and 1970s and which underpinned their lead in productivity and quality (Schonberger, 1982). Even though product technology still lags behind the U.S. in many fields, their production process organisation is unmatched anywhere. As Dosi (1981) comments, where other countries have taken their place within the international division of labour as given and hence have taken licensing (i.e. the institutionalisation of technological dependency) and inward investment as parameters, Japan has treated them as variables and has striven successfully to change them.

In semiconductors, after an initial success with transistors in the 1960s, considerable long-term R&D effort was devoted to catching up with the U.S. in a particular and well chosen range of chips (e.g. Dynamic Random Access Memories), in which it could take most advantage of its lead in high quality volume production (UNCTC, 1983, p.275). Since 1979, the Japanese firms have increasingly undercut U.S. and European producers with successive generations of products, while gradually moving 'up-market' to more advanced products such as microprocessors. For example, Japanese firms took over two thirds of the world market in 64K memories, 98% of the 256k market and are likely to dominate 1 megabit memories. This procedure of establishing dominance first in mass produced goods is also apparent in the computer peripherals industry where they have become dominant in printers.

The role of the Ministry of International Trade and Industry (MITI) is often cited as a key element in the rise of Japanese industry, though as we shall see other forms of intervention have probably had more effect in information technology. MITI has arranged low cost finance, export incentives, tax credits for co-operative R&D and investment and, above all, it has facilitated the transfer of technology within and between Japanese firms. Basic technology advances are considered common intellectual property but product developments are not shared and here there has been vigorous competition within the domestic economy even when the national economy has been protected by tariff barriers. However, although MITI is responsible for overall information technology policy, the large procurement and R&D budgets of NTT (Nippon Telephone and Telegraph) gives it more influence and power than MITI - and the fruits of its R&D efforts are quickly disseminated to

the IT firms (McKinsey, 1983).
Synergies between sectors have again had a
positive effect within Japanese electronics. In this
case it has been most notable between the consumer
electronics industry (in which Japanese firms
rapidly achieved world dominance in the 1960s and
1970s) and the components industry (Morgan and Sayer,
1983). Now as consumer markets begin to saturate, and
as the semiconductor industry increasingly converges
towards the computer systems industry through
increased miniaturisation of integrated circuits,
the major firms are moving into computers, office
equipment and industrial automation. The large
vertically integrated nature of firms like Hitachi
facilitates this process and also has the advantage
of co-ordinating design between components and the
systems products in which they are incorporated. The
obstacles to success, particularly in the shape of
IBM, are formidable. However, as a result of a co-
ordinated strategy in which MITI sponsored three
competing computer groups rather than a single
'national champion', Japanese firms have cut IBM's
share of Japan's annual domestic sales from 60% in
the 1960s to 25% now. This is IBM's lowest
penetration of any major country (Economist,
4.5.1985).
The Japanese have also benefited from cheap long
term loans supported by the government and hence have
been able to continue investment during slumps which
temporarily incapacitated overseas rivals, thereby
making themselves better placed to take advantage of
the succeeding boom. Lastly, they have used extensive
formal and informal methods to block inward
investment and imports, but unlike many countries
they have used the space created by such means to
build up their own capability. Such points illustrate
once again the particular social forms of capital, in
this case a symbiotic relationship with the state,
make a significant difference to inter-capitalist
competition and to patterns of uneven development
(Cf. Morgan and Sayer, 1983 and Sayer, 1985a).

Internationalisation from the Core Countries
We have laid stress on the home bases of the
multinationals (i) because these serve as launching
pads for their exports and foreign direct investment
and (ii) because MNCs often enjoy preferential
treatment in their home markets. While size of the
home market makes a difference to such firms the
larger companies cannot expect to survive and to

continue to accumulate capital while remaining
within the confines of even large domestic markets.
Internationalisation, through exports (internationalisation of commodities) or foreign direct
investment (FDI) (internationalisation of productive
capital) become imperatives. Again, it is capital
accumulation and not technological expansion or
progress which primarily drives this process:
burgeoning R&D outlays have to be amortised and not
even the massive American market is large enough to
permit this for many products, which is one reason
why internationalisation is so vigorous.

The most striking thing about global patterns of
trade and FDI in this industry, as in most others, is
the overwhelming dominance of the advanced economies
as destinations - and this, even allowing for the
runaway industry and 'Newly Industrialising Country'
phenomenon. This reflects the fact that the most
common reason for FDI is market penetration. (Note
however, this need not necessarily be counterposed to
cost minimisation as a reason, but could be quite
consistent with it.) Inevitably, given the
technological sophistication of the products, rich,
advanced markets are dominant, but why should firms
need to locate within them to service them? One
obvious reason is to avoid protectionism, whether
actual or threatened, and certainly this has been
important; for example the EEC has a 17% tariff on
semiconductor imports which has had the dual effect
of encouraging overseas producers to locate in the
EEC and discouraging European producers from
importing semiconductors back from offshore
production bases (1). Even where there are no
tariffs, voluntary export restrictions or quotas,
firms may locate to win approval from host
governments, who may be possible customers. It should
also be remembered that they may qualify for generous
state aid, even though this enables them to compete
more effectively with indigenous firms. For example,
the much vaunted American and Japanese electronics
companies of Scotland's 'Silicon Glen' have obtained
far more state aid than many British companies.

But there is also a further reason for FDI which
is often overlooked, especially on the Left with its
obsession with mass production consumer industries,
and this has to do with buyer-seller relationships.
It is important that much of the electronics industry
sells primarily to industrial and commercial
customers (e.g. a semiconductor firm to a
telecommunications firm). Many of the products are
fairly customised and therefore their design and

development requires close interaction in marketing
and development between the seller and buyer. Even
standard products like mass produced microprocessors
require similar interaction to help customers define
their needs, to help them learn how to use the
products, and to provide after sales customer service
for maintenance and subsequent adaptive work and
enhancement. This kind of relationship is common in
both hardware and software and revenue from the
service may be as or more important than that from
the product in some cases. A further consequence of
this is that R&D, especially Development, tends to
merge with marketing and service (innovations often
originate from the customer - see Von Hippel, 1977).
As a result, some development operations, as well as
production, need to be co-located with marketing and
servicing, at least in part, though not necessarily
within the same region of the host country. In the
case of information technology, markets are heavily
spatially concentrated in the metropolitan regions
of the advanced countries where information
processing operations of firms and public
institutions are chiefly located.

Large, vertically integrated companies regard
foreign suppliers with caution (though less so in the
case of European buyers). If they are to commit
themselves to using the suppliers' products they
expect evidence that the supplier has the production
capacity and/or second-sourcing arrangements to meet
the demand on time. This has prompted some ambitious
European companies to buy or build production bases
in the U.S. in order to create a sufficiently high
profile for winning orders from large indigenous
firms, even though this might not have been the
lowest cost location. Moreover, at the present time,
as leading firms try to reduce inventories and
tighten up on ordering, proximity to suppliers is
becoming more important (Sayer, 1985a).

In some cases, FDI for market penetration may
involve a company selling advanced products which it
already sells in its domestic market but which are
new to the host market. This is reminiscent of the
product cycle pattern (Vernon, 1966). In others, the
firm may be involved in a 'bootstrapping' strategy
where it attempts to gain access to a market more
advanced than at home in order to enable it to move
up-market towards 'state-of-the-art' products. This
latter strategy has figured prominently in the
reasons for investments by European firms in the
U.S., although the sheer size of the American market
(half the world market for computers) is obviously

21

important too. For example, the British chip firm
INMOS sold 80% of its output in the U.S. and its
chief executive commented - '... many UK customers
complain that its products are too advanced for their
needs. Inmos is an irrelevance to European industry
at this stage' (quoted in de Jonquieres, 1985a) (2).
In fact, despite the fact that INMOS was started,
with heavy state backing, as a reply to foreign
dominance in chips, over half its investment and
employment has been located in America, the primary
reason being the market, and a secondary one the
greater availability of leading skills in silicon
technology. There have been many other similar cases
where such moves have been considered essential for
success.

As we noted earlier, cost minimisation motives
are not necessarily inconsistent with a choice of
advanced country locations, but there are clearly
well known cases where certain labour intensive
production operations can be decentralised to
locations with cheap labour. This is most strikingly
the case with the 'runaway industries' in the Far
East and Central America. But it is also evident in
the particular choices of production locations by
inward investors attempting to gain access to the EEC
market. The more the production involves unskilled or
semi-skilled labour and shiftwork, the more likely
are areas with cheap labour (and usually regional
aid) to be chosen. This has been a major factor in
the 'success' of the British and Irish electronics
industries. Indeed cheap semi-skilled labour and
technicians in the UK are now being hailed as a
tremendous advantage in attracting manufacturing
activities to the UK, rather than to, say, Singapore
(ACT, 1985).

Such operations are increasingly integrated
into 'part-process' spatial divisions of labour (cf
Massey, 1984) spanning a whole continent or more;
that is, formerly largely separate and self-
contained foreign operations are specialised and
integrated into a supra-national production system.
(This is also sometimes termed 'complementation'.)
Other operations, particularly sales and service
branches, cannot be incorporated in this way and
still have to be replicated (or 'cloned' in Massey's
terms) for each region to ensure access to customers.

However, it is the 'runaway industries' which
have attracted most attention, although their
importance and significance has frequently been
exaggerated, most influentially in the 'New
International Division of Labour' thesis (Frobel et

al, 1980). Strictly speaking, this phenomenon is a subset of the more general case of FDI by advanced country firms in the developing or underdeveloped countries, namely that part which is aimed primarily at acquiring a low cost production site from which to export to third countries, usually rich ones. The first point to bear in mind is that it almost certainly constitutes only a minority of FDI in the Third World (Nayyar, 1978). McEwan argues, for instance, that multinational computer firms invest overseas mainly to penetrate their markets, both because of import and local content restrictions and because proximity to customers - particularly the state - is helpful. "Within Latin America, for example, US investment in manufacturing has become increasingly concentrated in Brazil and Mexico, the nations with large, rapidly growing markets, not the nations with the cheapest labour" (McEwan, 1985). Sometimes such markets may seem to be over-provided for by MNCs and indeed this is frequently the case as none wants to concede a market with major growth potential to competitors.

The runaway option (in the strict sense) is obviously only open to certain kinds of activity: mass produced, high value per unit weight, and involving a substantial element of low-skilled manual labour and lacking a need for intensive technical management or proximity to customers.

In semiconductors, the first runaway firm was Fairchild, which established a labour-intensive assembly plant in Hong Kong in 1962. A procession of American electronics firms followed, locating first in the Newly Industrialising Countries and later diffusing out to countries with still cheaper labour such as the Philippines and exporting the assembled products back to advanced country markets (Chang, 1971; Ernst, 1981; Rada, 1982). A smaller number of plants was set up in Latin America, particularly Mexico. These areas offered not only cheap female labour, but longer working hours, higher work intensity, scope for multiple shifts, dispensability and replaceability of workers, state repression of labour organisation, tax holidays and other benefits such as free repatriation of profits, and last but not least, state provision of infrastructure (Pearson and Elson, 1981). As with many developments widely attributed to naked market forces, state intervention played a major enabling role (Henderson, 1985).

Spectacular though the runaway industries are, there are many limits to their development. First, as already implied, the lack of local markets and technical skills restricts the kind of activities which can be decentralised. Second, wage differentials have narrowed between the NICs and the West. Third, worker resistance has frequently developed; for example the recent waves of (illegal) strikes in semiconductor plants in the Philippines (Electronics Times July, 1985). Fourth, political instability, again particularly in the Philippines, has threatened inward investment. Fifth, most of the runaways have been American, largely because US tariff structures (unlike European ones) are favourable to the re-importation of semiconductors. If all these factors affect the accumulation process as external, though pervasive contingencies, there is a sixth factor which is more endogenous to the process, that is the displacement of manual labour by automation and product innovation. As automated methods become more general, the attraction of cheap labour countries is diminishing (automation can enable one worker with two weeks' training to replace thirty manual assemblers with three months' training each (Rada, 1982)) and many new assembly plants are being located in the 'North'. Some firms, particularly the Japanese and 'captive' producers like IBM, have hardly used offshore manual assembly, preferring relatively automated production 'on-shore'. Furthermore, the amount of semiconductor production located in the advanced countries is also rising with the trend towards more customised chips produced in smaller batches for which the need for proximity and rapid response to customers outweighs the advantages of internationalised production systems.

There is also an emerging pattern of differentiation among the countries of the Far East, due to a combination of 'market forces' (i.e. unintended consequences of actions of individual capitals) and state intervention. There is a growing division between two of the Newly Industrialising Countries – Hong Kong and Singapore – and other countries like Malaysia, Philippines, Thailand and Indonesia. The wage difference between the first two countries and the U.S. is less than for the second set and, as already noted, becoming narrower, and the first two countries also have more skilled technical labour, partly as a result of improved state technical education. The net effect of these two factors is to produce an outward movement of low-skilled work on low value products to the cheapest labour countries and an upgrading of domestic activities and increased capital intensity in Hong Kong and Singapore. Indeed these two countries are now said to have the beginnings of self-generating industrial complexes in electronics, including equipment manufacturing and centres for designing components for local industrial customers (Henderson, 1985). This illustrates the more general point that although there is a clear spatial division of labour in the industry it should not be thought of as fixed but as evolving as industries and local economies change.

The internationalisation of the Japanese electronics industry deserves a separate discussion. As noted earlier, the behaviour of a subset of American multinationals has frequently been taken as the general model for the behaviour of international capital, rather than as the product of a particular form of capital within a specific conjuncture. Against this 'norm', the behaviour of the major Japanese companies appears as an aberration, though they conform to the imperatives of capital accumulation no less than the Americans. Their reputation as 'reluctant multinationals' owes much to the superior productivity which they can achieve within Japanese society but find difficult to match overseas, and secondly to the favourable relations with the Japanese state and financial capital. Hence their preference for exports over FDI and for mass market products which need less representation in local markets. Virtually their only reason for establishing production facilities in other developed countries has been to pre-empt protectionism and, in America, to monitor technology. Within Europe, language, relatively low wages and encouragement to inward investment appear to have been their reasons for choosing Britain and Ireland as their main bases. They have also made less use of the runaway option, as is indicated by the fact that their imports from the rest of the world (excluding the U.S. and Europe), were only 8% and 15% of those of the U.S. and Europe, respectively, from this source.

As we have seen, locationally specific advantages and disadvantages have had a marked effect on competitiveness within the industry. It has also been clear that nation states are significant actors in the competition. This is especially noticeable in the case of response to perceived threats from foreign competitors, be it British or EEC governments responding to American dominance or the U.S. government responding to the 'Japanese invasion'.

Three sets of relationships are involved each combining competition, conflict and cooperation - not only capital and labour, but relationships between capitals (of the same or different nationality) and between capitals (foreign or domestic) and governments. These last two relationships are further complicated by foreign direct investment, because 'domestic producers' can be taken to include foreign firms (e.g. ITT and IBM in Britain), while 'foreign industry' can include subsidiaries of domestic firms e.g. INMOS's U.S. plant. Given this tangle of contradictory relationships, alliances are bound to be temporary and opportunistic.

This is perhaps most evident in Europe where there have been repeated calls for EEC wide collaboration to prevent further American and Japanese encroachment and additional efforts by individual governments to mount regeneration programmes. Few of these have met with even partial success. Not only are the European companies competitors, but even though they may wish to collaborate to repel non-European competitors they are simultaneously driven in the short run to make deals with them in order to catch up with their technology and get access to overseas markets. In fact, they have frequently been more ready to do the latter than the former. For example, the British computer firm, ICL, has now come to rely on Fujitsu for the manufacture of 80% of its chips. In the case of the ESPRIT programme, specifically designed to resist American and Japanese dominance, IBM has managed to join the programme as a local producer, despite the protestations of many of its rivals. There have also been complaints about the 'problem' of imports from the Far East, even though some of the larger MNCs (like Philips) are actually involved in exporting to Europe from their own offshore subsidiaries.

Examples of Corporate Spatial Divisions of Labour

Corporate spatial divisions of labour reflect all these influences plus those deriving from the particular cost structures associated with designing, making and selling particular products. Perhaps the most striking examples of international production systems are to be found in the major American merchant semiconductor firms, such as Texas Instruments, Motorola and National Semiconductor. These locate most of their R&D in the U.S.,

particularly in Silicon Valley, in order to benefit
from close monitoring of the most advanced markets
and the local concentration of top scientific skills.
They also tend to have small R&D units in major
foreign markets, usually concentrating on development work in relation to major 'local' markets.
Manufacturing consists of two main stages, wafer
fabrication and assembly. The former is highly
capital intensive and requires a fair amount of
technical labour and consequently this tends to be
concentrated in the U.S.. High labour and land costs
in California have tended to drive out this activity
to other locations in the southern and western U.S.,
leaving mainly fabrication of the most advanced and
customised products in California. Fabrication
plants have also been set up in Europe, particularly
Scotland to get inside the EEC tariff, improve market
access, take advantage of the generous state
investment grants and to use cheaper semi-skilled and
skilled technical labour. As already noted, heavy use
has been made of Third World locations for labour-
intensive manual assembly, though a significant
amount of assembly of smaller batch orders is still
done in the U.S., even in California, using cheap,
mainly female Hispanic immigrant labour, and within
Europe, especially in the Republic of Ireland. With
the rise of automated methods of assembly, the firms
are locating new assembly operations in the U.S., as
the major captive producers like IBM and AT&T have
always done (Rada, 1982; UNCTC, 1983). Finally,
marketing and customer support - both labour-
intensive activities requiring skilled labour and
good access to customers - are located in all the
major markets.
    No firm in the IT industry can compare with IBM,
with its massive domination of the world computer
industry; to other computer firms IBM is not so much
a competitor as the environment. This extraordinary
strength is reflected in its vertical integration and
in the extent and complexity of its global
organisation. Globally, its dominance of virtually
every national market has posed the danger of
nationalisation or exclusion by host governments and
this has affected IBM's locational strategy. In
response, it has developed an elaborate system of
complementation i.e. a network of manufacturing
plants each specialising in just a few parts of
computer systems and linked by large cross-border
flows of materials and products. No one computer
system is made entirely within one country (which
reduces the threat of nationalisation) and imports

and exports to and from each country are kept roughly
in balance so that the company cannot be charged with
upsetting the balance of payments. For example, the
Greenock plant in Scotland now assembles
input/output devices and terminals. Havant, in
Hampshire assembles and tests various peripheral
equipment, including disk storage systems. Boca
Raton, Florida, assembles certain mainframes and
Personal Computers. Bordeaux does printed circuit
board assembly and test, Montpelier manufactures
telecommunications equipment and Virnecate near
Milan assembles large minicomputers. Components are
manufactured in Burlington, Vermont and East
Fishkill, N.Y., Toronto and many other locations (3).
And so the list goes on - but it is worth adding that
the company has made no use of 'runaway' plants in
the Far East.

The huge R&D budget, exceeding that of the U.S.
space program, is also distributed across a large
number of specialised but linked operations. Perhaps
significantly, research in some of the newer fields
such as distributed data processing has been
decentralised to locations in the western U.S., such
as Boulder, Colorado. Examples of overseas R&D
activities are Hursley, UK (disk storage systems) and
Nice, France (satellite communications and fibre
optics). According to Bakis (1980), each of these
overseas operations is paralleled by an equivalent
specialist lab in the U.S. with which it cooperates.
Again, the reasons for this decentralisation are
partly political, but given the company's size, the
need to relate R&D to customers' evolving needs and
the global shortage of skilled labour it presumably
cannot afford not to have foreign research labs.

Given the importance of marketing and customer
support in the industry IBM has scattered numerous
service operations in all its major markets, the more
routine activities forming a 'cloning' structure
(Massey, 1984) similar to a central place network,
the more specialised formed on a 'complementation'
basis. While this kind of activity has tended to be
neglected in industrial location studies, it is an
essential element of a computer firm's competitiveness.

Enormous amounts of information are generated
both by the need for close, continual interaction
with customers and by the complex product range and
corporate spatial divison of labour. This in itself
requires a network of internal information
processing centres for coordinating activities
around the world; e.g. Cosham, near Southampton,

Uithoorn in the Netherlands and Boulder, Colorado are just three elements of a vast network of data processing centres linked by the most advanced telecommunications. As Bakis shows, this network enables the company to circulate virtually every operation in the world with relevant information from any part of the system within 24 hours, thereby reducing response times and greatly increasing the utilisation and turnover of capital.

Finally, Northern Telecom, a leading telecommunications firm based in Canada, provides an illustration of the lower degree of internationalisation in this sector. As is increasingly the case in telecommunications, the company manufactures chips and computer systems for its own use. It has 47,000 employees spread across a large number of sites; there are 27 manufacturing plants in Canada, 15 in the U.S., one each in the Republic of Ireland, Brazil, UK and two in Malaysia, and 25 R&D units, some of them incorporated with manufacturing, indicating the customised nature of many of the products, particularly software. Twenty-one of these are in Canada, 15 in the U.S., and one in Maidenhead, UK. Most of the foreign investment has been directed at supplying the less restricted types of foreign market, particularly business telecommunications and information processing. U.S. revenues were 64.5% of total revenues, Canadian 31.5%, leaving only 4% from outside North America (Northern Telecom, 1984).

Britain's Place Within the International Division of Labour in Electronics

Before concluding, we will now briefly review Britain's place within the IDL in electronics. (By Britain we mean any industry located within Britain whatever its ownership, but excluding British overseas investment.) Views on this topic are remarkably polarised, ranging from that of the government, which sees it as healthy and strongly placed, to that of the National Economic Development Council which sees it as in 'relative decline' and facing 'impending crisis' (NEDC, 1982). Yet there are some good reasons behind this polarisation, for the position of Britain is certainly paradoxical.

On the one hand, there is a rapidly deteriorating electronics trade deficit in Britain; a £150m trade surplus in 1975 had become a deficit of £2.6bn by 1983 and this deficit is forecast by Cambridge Econometrics to grow to £8.3bn by 1993 (de Jonquieres, 1985b). On the other hand, Britain has had unrivalled success in attracting inward investment, being the most favoured host country in Europe for U.S. and Japanese firms. Normally one would expect this to have had an import substitution effect. However, it has clearly not done so. Partly this is because FDI in production in Britain has not kept up with demand, and many of the activities located here are primarily marketing arms of companies which suck in more imports. Also, as we have seen, indigenous companies have frequently failed to penetrate new markets or ceded them to foreign competitors. While many submarkets are technically less sophisticated than in the U.S., there are some which are attractive; in particular the consumer electronics market is seen as the most advanced in Europe, the most liberalised in telecommunications, and the most open in semiconductors and computers.

Furthermore, relative to many other European countries, Britain offers low wages, not just for semi-skilled labour but for technicians and technologists, especially in software, though there are skill shortages in this field. However, such advantages are likely to be appropriated by foreign firms at the expense of indigenous companies.

Inward investment has been encouraged by the most indiscriminate 'open door' policy in the EEC (with the possible exception of the Irish Republic). While the Thatcher administration sees this in pro-market terms as an example and stimulus to British capital, the heads of GEC and ICL have opposed the policy of offering foreign competitors subsidies to compete on their doorstep. So while the rise in output in the British electronics industry far exceeds that of most other domestic industries, it still lags seriously behind that of the leading firms and countries. And as the NEDC has warned, it is not sufficient to remain competitive in this rapidly changing industry.

Conclusions

Although this has been an empirical study of a particular industry it is possible to draw out some theoretical implications of more general relevance, especially for the radical literature on the IDL and on spatial divisions of labour.

First, and least contentiously, it is certainly true that firms respond to pre-existing patterns of uneven development, not just in terms of where they locate their various activities but in terms of how

they divide up their activities in the first place
(Massey, 1974). This was clear, for example, in the
runaway industries, and again in their possible
decline. But the radical literature seems to be
transfixed with labour as the only significant
'locational factor'. At the same time the effects of
the need for market access have been treated as
insignificant. This is perhaps because at the
subnational level, where the spatial division of
labour concept has primarily been used, the market
seems pervasive rather than locationally-specific.
And this tendency is further reinforced in the
radical literature by the common implicit assumption
that all markets are mass consumer markets and hence
dispersed, ignoring specialised industrial markets
which are frequently quite localised. What has tended
to be forgotten - and the omission becomes much
clearer at the international level - is that the
spatial division of activities is affected not only
by labour considerations but by market variations.
Traditionally, before the work of researchers like
Hymer, the reverse tendency was common, namely, an
underestimation of the extent to which multinationals could divorce their production locations from
their markets. We have now to avoid going to the
equally unsatisfactory opposite extreme.

Second, and also apparently uncontentiously, it
is clear that firms react back on the localities
within which they operate. Yet again, this
relationship tends to be seen rather restrictively,
purely in terms of the effects of firms on local
labour markets. But they also affect their customers
and sometimes induce local suppliers (IBM, 1984).
Now in the traditional regional literature (e.g.
growth centre theory) this latter possibility was of
course given excessive weight; studies of MNC's
branch plants repeatedly showed that they made far
less use of local sourcing than did smaller
indigenous firms. Yet there is a danger in over-
reacting against the growth centre notion, firstly
(again) because of the obsession with mass production
at the expense of other activities more likely to
make use of local sourcing (e.g. applications
software) and secondly because of an ignorance of
recent changes in favoured methods of production
which follow Japanese practice in requiring close
relationships with suppliers and regular deliveries.
For example, both at its Montpelier (France) and Boca
Raton (Florida) Personal Computer plant, IBM has
encouraged several major suppliers to relocate close
by in order to coordinate production more tightly

(Business Week, 3.10.1983; Bakis, 1980).

A third problem with the radical literature which the empirical research exposes is the tendency to focus on lower skilled labour at the expense of the top end of corporate hierarchies, so that R&D, marketing and service work are treated as uninteresting black boxes. Spatially, this is reflected in a disproportionate interest in the periphery relative to the centre. In the case of 'high technology' industry, if we look at these 'top end' activities more closely, particularly their buyer-seller relationships, it becomes clearer why proximity to major customers is essential and easier to understand one of the main reasons for investment in advanced countries. There is also a reluctance to accept that certain top technical skills are in short supply in many radical accounts. Yet they certainly exist in the electronics industry and indeed they have obliged leading firms to disperse their R&D activities much more widely than might otherwise have been the case. (This internationalisation of R&D is also influenced by the need to interact with major industrial customers in different national markets).

A fourth problem relating closely to the above concerns possibilities for change in spatial divisions of labour, whether at the international or subnational levels. In the traditional, non-structuralist literature, corporate activities are conceptualised in abstraction from the larger structures of interdependent activities of which they are part. This permits the common fallacy that many areas can simultaneously 'clone Silicon Valley', as if there were enough 'top end' activities to go round every development agency that wanted them. But there is also a danger, possibly deriving from some residues of dependency theory, of going to the opposite extreme, where it appears that spatial divisions of labour are fixed, or that if change does occur, it can only involve individual places or firms changing places within that fixed structure. Clearly, as we have seen, some are changing their place (Japan, Hong Kong, etc), but while they are undoubtedly changing within the wider structure of the IDL they are also helping to change the nature of that structure too. It is not like changing seats on a bus, for the nature of the seats and the transport are changing, though without collapsing into an unstructured mass in which individuals can do anything, regardless of others. (4)

A further implication of this again concerns the effect of firms on localities. It was clear that firms reproduce particular local labour market characteristics (e.g. low paid labour in the Far Eastern countries) and that often the continuation of their activities in those localities is dependent on the reproduction of those conditions. But those conditions are frequently changed, whether intentionally as in the case of training labour where there are skill shortages, or inadvertently, as in the case of the upward drift of wages in high growth areas such as Silicon Valley or Hong Kong.

Our fifth point concerns a tendency to conceptualise corporate structures as simple hierarchies as in the Hymer conception quoted earlier. While there is obviously a hierarchical element, there is a danger of exaggerating the centralisation of power and focusing on vertical divisions at the expense of horizontal ones. Massey's distinctions between different kinds of corporate spatial division of labour, such as 'part-process' or 'cloning' are helpful in moving beyond this, but ultimately we have to pay attention to the actual activities of companies instead of assuming them to be nothing more than places within a hierarchical conception-execution continuum. Again, this relates back to the need to examine all parts of corporate hierarchies and not just their lower echelons.

Sixthly, it is abundantly clear from the study of the electronics industry that the character of national economies and their economic policies has had and will continue to have a major influence upon the IDL, though there is little acknowledgement of this in the theoretical literature.

Finally, let us return to Hymer's provocative hypothesis about multinationals and the internation-al division of labour: are the economic characteristics of countries and regions correspond-ing increasingly with their role with respect to the local operations of multinational companies? The answer must be yes but far less than Hymer implied, and certainly with many complications. The main caveat is the limited and uneven (though increasing) amount of foreign direct investment and the effect of nation states as both constraints upon and actors within the IDL. While there is a convergence between the character of corporate activities and the local economies in which they operate, the former are not simply hierarchically organised, and both the activities and the local contexts are more diverse than the literature on the IDL appears to recognise. And lastly, given the importance of markets and national initiatives in this and other industries,

the IDL is overwhelmingly skewed towards the advanced
countries of the world with the runaway industries as
an exception rather than as the norm.

Notes
1. The tariff for complete boards of chips,
however, is only 5.3%, so the incentive to assemble
'systems products' inside Europe is less than for
semiconductors themselves.
2. Similarly, the head of Philips semiconductor planning said "Europe simply does not provide
the 'market pull' needed to create world product
standards" (de Jonquieres, 1985a).
3. Most of this information on IBM comes from
Bakis (1980).
4. In other words, in Giddens' terms, we are
proposing a 'structurationist', instead of a
structuralist, approach (Giddens, 1979).

References
Applied Computer Techniques (ACT), (1985) House of
Lords Select Committee on Overseas Trade, Minutes of
Evidence
Bakis, H. (1980) 'A Case Study of IBM's Global Data
Network', paper presented to the 24th International
Geographical Congress, Tokyo, August
Business Week (1983) 'How the PC Project Changed the
way IBM Thinks', 3.10.1983
Chang, Y.S. (1971) 'The Transfer of Technology: the
Economics of Offshore Assembly: the case of
Semiconductor Industry', United Nations Institute
for Training and Research, No. 11, New York
de Jonquieres, G. (1985a) 'The Harsh Imperatives of
Survival', Financial Times, 24th June
de Jonquieres, G. (1985b) 'Electronics Trade Deficit
Will Deepen', Financial Times, 18th February
Economist, (1985) 'IBM in the Ackers Era', 4th May
1985
Electronic Times (1985) 'Violence Erupts on
Philippine Picket Line', 4th July 1985
Elson, D. & Pearson, R. (1981) 'Nimble Fingers Make
Cheap Workers: an Analysis of Women's Employment in
Third World Export Manufacturing', Feminist Review,
7, 87-107
Ernst, D. (1981) The Global Race in Microelectronics,
Verlag, Frankfurt
Frobel, F., Heinrichs, J. and Kreye, O. (1980) The
New International Division of Labour, Cambridge
University Press, Cambridge

Giddens, A. (1979) Central Problems in Social Theory, Macmillan, London
Henderson, J. (1985) 'The New International Division of Labour and American Semiconductor Production in South-East Asia' in D. Watts, C. Dixon, and D. Drakakis-Smith (eds) Multinational Companies and the Third World, Croom Helm, London
IBM (1984) IBM UK, North Harbour, Portsmouth
Hymer, S.H. (1975) 'The Multinational Corporation and the Law of Uneven Development' in H. Radice (ed.), International Firms and Modern Imperialism, Penguin Books, Harmondsworth, 37-62
MacEwan, A. (1985) 'Unstable Empire: US Business in the International Economy', Institute of Development Studies Bulletin, 16, (1), 40-6
McKinsey and Co. (1983) A Call to Action: the European Information Technology Industry, Commission of the European Community, Brussels
Massey, D. (1984) Spatial Divisions of Labour, Macmillan, London
Morgan, K. and Sayer, A. (1983) 'The International Electronics Industry and Regional Development in Britain', WP-34, Urban and Regional Studies, University of Sussex, Brighton
National Economic Development Council (1982) Crisis in the IT Industry, NEDC, London
Nayyar, D. (1978) 'Transnational Corporations and Manufactured Exports from Poor Countries', Economic Journal, 88, 59-84
Northern Telecom Ltd. (1984) Annual Report, Mississauga, Ontario
Rada, J. (1982) 'Structure and Behaviour of the Semiconductor Industry', mimeo, United Nations Center on Transnational Corporations, New York
Sayer, A. (1985a) 'Industry and Space: a Sympathetic Critique of Radical Research', Environment and Planning D: Society and Space, 3, 3-29
Sayer, A. (1985b) 'Industrial Location on a World Scale: the Case of the Semiconductor Industry' in A.J. Scott and M. Storper, (eds) The Geographical Anatomy of Industrial Capitalism: Production, Work and Territory, Allen and Unwin
Schonberger, R.J. (1982) Japanese Management Techniques: Nine Hidden Lessons in Simplicity, Free Press, New York
Sciberras, E., Swords-Isherwood, N. and Senker, P. (1978) 'Competition, Technical Change and Manpower in Electronic Capital Equipment: A Study of the UK Minicomputer Industry', Science Policy Research Unit Occ. Paper No. 8, University of Sussex
Soete, L. and Dosi, G. (1983) Technology and

35

Employment in the Electronics Industry, Frances Pinter, London

UNCTC, (1983) *Transnational Corporations in the International Semiconductor Industry*, United Nations Center on Transnational Corporations, New York

Vernon, R. (1966) 'International Investment and International Trade in the Product Life Cycle', *Quarterly Journal of Economics*, 80, 190–207

von Hippel, E. (1977) 'The Dominant Role of the User in Semiconductor and Electronic Subassembly Process Innovation', *IEEE Transactions on Engineering Management*, May, 60–71

Chapter Three

THE GROWTH AND INTERNATIONALISATION OF THE AMERICAN SEMICONDUCTOR INDUSTRY: LABOUR PROCESSES AND THE CHANGING SPATIAL ORGANISATION OF PRODUCTION

Jeffrey Henderson and A.J. Scott

Introduction

One of the most far reaching features of industrial production in the contemporary world system is its tendency towards reorganisation at a global scale. The foundations and motor forces of this reorganisation arise within what has come to be known as the 'new international division of labour' (Frobel et al 1980, Ernst 1980). These developments, which have emerged predominantly in the last quarter-century, are beginning to have important repercussions on urban and regional development (cf. Hymer 1979, Frank 1980, 1981; Cohen 1981 and Hoogvelt 1982).

In this paper we seek to elaborate an understanding of these developments by means of a conceptual and empirical investigation of a paradigmatic case: the American semiconductor industry. We show how the industry evolved in the post World War II period and how its peculiar pattern of evolution at this time created a set of distinctive spatial outcomes. We continue by showing how the 'creative destruction of capitalism' (Schumpeter) tended to dissolve the industry's spatial focus, resulting ultimately in its dispersal across the globe. We pay particular attention to the effects of this development pattern in the United States, South-East Asia and Scotland.

We see our work as a contribution to an ongoing debate about a series of important social, economic and spatial questions, and in particular to the whole question of the meaning and logic of the new international division of labour. We attempt to show that any adequate understanding of the new international division of labour necessitates attention to (a) the structure of capitalist commodity production at large, (b) the central dynamics of labour processes in the context of the contradictory capital-labour relation, (c) the internal and external organisation of industrial production, (d) the spatial consequences of these phenomena and (e) the way this ensemble of relations changes over time.

We wish to stress at the outset that we see our work in this paper as being, of necessity, provisional. We attempt to identify some of the main theoretical coordinates which we feel must underlie any effective attempt to conceptualise the globalisation of semiconductor production. While this paper draws on our own fieldwork in California, Hong Kong and Scotland, we must mark the general absence, as yet, of highly developed bodies of data and case studies of particular firms or production complexes (though the work of Saxenian 1981, Lim 1978 and Lim and Pang 1982, constitute exceptions here). For these reasons, the present paper should be taken as an interim assessment; i.e. an attempt to identify emergent patterns and tendencies, rather than the elaboration of definitive empirical and theoretical conclusions.

Before we reach the body of the paper, it is necessary to describe some of the semiconductor industry's products and production processes.

The Industry and its Product. Semiconductors are the central component in the transmission, reception and amplification of electronic signals. As such they are essential to telecommunications, data storage, information retrieval and manipulation (computers), home entertainment (TV, radio, hi-fi, video etc), medical science, military hardware, aerospace etc. They are in short, the base component of machines whose absence would render contemporary urban-industrial society unworkable, and social life within it, almost unthinkable.

The precursor of modern semiconductors was the vacuum tube. This device, invented in England shortly after the turn of the century, consisted of from two to five or more electrodes surrounded by a glass tube. The tube was sealed to a metallic base and a vacuum was induced inside it so as to improve conductivity. Following the development of wireless telegraphy, the vacuum tube (or 'valve'), became central to the development of all electronics for much of the following half century.

Though it could be easily and cheaply manufactured, the vacuum tube was fragile, bulky, relatively slow in transmitting electronic signals, and because of its incandescent filaments, consumed relatively large amounts of power. By the middle of the 1950s, partly because of these disadvantages, the vacuum tube began to be replaced by a form of semiconductor on which the subsequent 'microelectronics revolution' was to be based. The new semiconductors emerged technologically from developments in solid-state physics during the 1920s and 1930s, and their first generation was constituted by the transistor.

The transistor had been invented in the United States shortly after World War II. It was first made in commercial quantity by etching circuits on germanium (and subsequently silicon) wafers by means of electrolyte jets. Individual circuits were then cut from the wafer, and manually (using microscopes and soldering irons) attached to wires (which facilitated their connection to other transistors), and then sealed (bonded) in a casing of non-conducting material.

The first generation transistors were relatively unreliable, could cope with only low frequency signals, and the fabricating process produced very low yields. By the late 1950s, a process of wafer-fabrication - the 'planar' process - had been developed which was subsequently to help revolutionise once more, semiconductor technology. The planar process used photolithography to transfer the circuits traced on 'masks', and etch them onto the surface of the silicon wafer. The wafer was then subjected to a chemical diffusion process which transferred certain impurities possessing particular electronic qualities to the surface of the wafer. Though the planar process helped to increase yields (i.e. successfully fabricated wafers and circuits) and improve reliability, the assembly process (i.e. wire attachment and bonding) remained unaltered.

The revolutionary character of the planar process arose from the fact that it technologically allowed not only for the possibility of producing many separate circuits from the same silicon wafer, but also for creating many circuits on the same wafer fragment. As a result it constituted an essential precondition for the development of those semiconductors that are the principal building blocks of electronic machines to this day: integrated circuits.

As its name suggests, the integrated circuit ('silicon chip') consists of a multiplicity of circuits, etched on a single fragment of silicon.

Integrated circuits have relatively low power consumption, are highly reliable, cheap to produce, and most importantly are tiny devices which perform more quickly the functions of earlier generations of electronic machines.

Since their first commercial development in the early 1960s, integrated circuit technology has become increasingly sophisticated in two senses. Firstly, the number of circuits that can be carried by a single chip and hence the number of functions that the integrated circuit can perform have expanded enormously. With first 'large scale' (LSI) and then 'very large scale' integration (VLSI) in the 1970s, each chip in the latest generation of integrated circuits now contains in excess of 100,000 microscopic transistors. Secondly the range of functions that the integrated circuit can perform has advanced significantly. There are now five forms of semiconductor technology ranging from the simplest, linear technology (embodied in the form of transistors and now generally referred to as 'discrete' devices and used, for instance, in telephone systems) to the most advanced microprocessors which are able to manipulate data and hence perform the functions of computers. Although the continued development of integrated circuit technology has had enormous implications for research and development work and for the masking and diffusion of silicon wafers, until recently it had few consequences for the assembly and final testing of semiconductors. Only in the last few years, partly in response to the evolution of VLSI technology, have assembly processes begun to be automated, and final testing begun to be transferred to the control of lasers and computers, (cf. Borrus et al 1982, Wilson et al 1980).

In view of this brief but necessary excursion through some of the technological developments and terminology of the industry, we now turn to the body of the paper. We begin by raising some of the theoretical and methodological issues that inform our discussion.

Some Theoretical and Methodological Preliminaries

In addition to its intrinsic empirical interest and significance, the present study may be seen as a contribution to a series of on-going theoretical debates about location theory, the spatial division of labour and territorial development. Our purpose is to show, first of all, how the American semiconductor industry became concentrated in the Santa Clara Valley of California and then, as it grew, began to disperse at a rapid rate to a wide variety of offshore locations. But we are also at pains, second of all, to make explicit how these geographical developments and shifts relate to a very general system of dynamics within capitalist labour processes. These dynamics can only be comprehended as they themselves grow out of the (non-spatial) relations of production of capitalist society at large.

The elementary starting point for any examination of relations of production in capitalism is, of course, a notion of the production system as a domain of labour organised and overseen by private firms. Firms supply materials and equipment (capital), hire workers at a wage, set in motion a labour process, and then ensure that the outputs that are produced in this process are sold at an appropriate price. Among other things, this price must be such as to secure for firms at least a normal profit on all capital advanced. In capitalism, the survival of firms depends critically upon their ability to earn normal profits or better. This ability, moreover, also depends on their continued cost-effectiveness relative to the competition (even where the competition is highly imperfect). This signifies at once that capitalist firms have no option but to press down more or less continuously on all of those expenses that otherwise might make inroads into their profitability. Thus in capitalism, firms typically and unceasingly seek to improve their productivity, to save on capital expenditures, and to hold back labour costs, where labour costs include not just wages but also fringe benefits, safety standards, internal environmental conditions, and so on.

This imperative constitutes the basic motor of the spatial theory that we develop in the sections that follow. It is the overarching condition within which a series of geographically critical manoeuvres, options, strategies, decisions, etc. on the part of both capital and labour are played out. It stimulates technological improvements, and it forces firms into a series of collisions and/or accommodations with their workers. At the same time, the bargaining power of the latter is constantly under threat of dissipation by reason of those same technological improvements which in various ways deskill and devalue the labour force (cf. Braverman 1974, Edwards 1979). All of this results in the appearance of the phenomenon of especially rapid technical innovation and change in sectors of production where the next increment of productivity gain can fairly consistently be extracted out of existing theoretical, technical, mechanical, and social arrangements. For the last two decades or so, the American semiconductor industry has had precisely this character.

As we shall see, the logic of these processes leads to a variety of different socio-political outcomes within the industry itself as well as in the human communities that grow up in its shadow and which provide it with its main labour force. Most importantly of all for present purposes, this logic induces a series of highly significant organisational and technical responses within the industry. On the one hand, and under specifiable concrete conditions, it may lead to rapid vertical and horizontal disintegration of labour processes within the industry, so that the whole sector becomes predominantly a complex of many specialised relatively small-scale producers with very detailed interlinkages. This leads at once to an observable tendency for producers to congregate together in geographical space. On the other hand it may encourage a reversal of these very same tendencies. Thus, we may also observe a countervailing trend to significant scale increases within producing units along with an underlying process of capital-deepening and/or rapid standardisation of labour tasks and outputs. As this latter trend becomes more and more dominant so firms grow in size while at the same time their several internal functions become more markedly spatially specialised. In this way the multi-establishment firm makes its appearance as a peculiar organisational form, involving <u>spatial disintegration, but functional integration</u> of labour processes. As we show, all of these tendencies have been observable at various stages in the development of the semiconductor industry.

Quite obviously the theoretical notions adumbrated above are nothing more than a few casual and selective pointers to a more careful examination of the logic of modern capitalist production systems. Their importance for present purposes is not just that they help to set the scene for the analytical specifics that follow (though they most certainly <u>are</u> useful in this regard); but more significantly, they are an entry point into a crucial methodological issue that must be resolved before a viable and effective location theory can be established. This issue hinges around the contrasting analytical problems and tasks raised by examination of (a) the internal necessary relations of any social mechanism and (b) the external contingent relations of the same mechanism (cf. Sayer 1982). It is especially important to raise this issue in the context of an empirical study of the semiconductor industry, because much of the literature on the industry has hitherto (to its cost) glossed over precisely this important distinction.

By a system of internal necessary relations we mean relations whose real existence is a sine qua non for the continued existence of a given set of phenomena. By the same token, an understanding of these relations is a necessary preliminary to any understanding of the phenomena they link together. In slave society, for example, we can only comprehend what a master is in the context of knowledge that this phenomenon is intrinsically related to another phenomenon, the slave. In contrast to this idea, a system of external contingent relations constitutes a set of relations that may very well have significant implications for the ways in which given phenomena operate, but which nonetheless do not constitute their intrinsic conditions of existence. This implies at once that these contingent conditions must by their nature remain at a secondary explanatory level in any attempted theoretical account. For example, the precise form of the institution of slavery (in any given historical realisation) may well be modified by the circumstance that there is a powerful democracy next door. But still this circumstance is not central to the generalised social logic of slavery. In brief, we can only understand systems of social phenomena as intertwined webs of necessary and contingent relations, dominated by the analytical priority of the former.

So it is with location theory in general and the spatial structure of the semiconductor industry in particular. We must understand the industry as being, in the first instance, a sector of production in capitalism and like any other sector, subject to the overarching pressures and dynamics of the capital-labour relation. This relation (as part of its own internally necessary operation) is endemically associated with such phenomena as accumulation, technical change, actual or potential labour conflict, and so on, as alluded to earlier in this section. At the same time, the dynamics of the capital-labour relation are projected out, as it

43

were, into specific social and spatial contexts, and these contexts <u>channel</u> the precise ways in which the relation is realised in specific circumstances without, however, modifying its intrinsic nature. They are in short, contingent (even if empirically important) circumstances.

In the semiconductor industry in particular, a number of contingent historical and geographical processes and events have helped shape the precise empirical form which the structure of the industry has taken on. Of these, some are 'accidents', such as the decision of one of the industry's founders to choose Santa Clara County out of the scores or even hundreds of alternative locations that might have been selected. Others are related to geographical conditions in earlier phases of development of the capitalist world system (cf. Wallerstein 1979, 1983). The existence of pools of cheap labour in South-East Asia just at a time when certain segments of the semiconductor industry were able to internationalise on a large scale, and the emergence of much of the central valley of Scotland as a 'Special Development Area' offering relatively cheap production sites within the EEC tariff barrier, are cases in point. All of these circumstances have had a profound and enduring impact on the locational structure of the industry, and accordingly we take them very seriously in our discussion below. However, we should not conflate them with the central necessary dynamics of the industry. To repeat, our objective is a locational analysis that is sensitive to the different theoretical levels and concepts brought into play by a conception of spatial process as an interpenetrating constellation of necessary and contingent relations. In this way it is our hope to lay the foundations for a more effective understanding of the spatial political economy of what is, after all, an extraordinarily important industry.

We find it necessary to press these points home precisely because the semiconductor industry (like many other industries) has so frequently been treated by location theorists in ways that we consider to be incoherent, i.e. in ways that take contingencies as basic explanatory categories, and that overlook entirely the essential central driving mechanisms of the industry. Our point of departure, by contrast, is a conception of the semiconductor industry as an element of the production system of capitalism at large, and therefore a conception that insists upon the structuring of the organisational characteristics of the industry by the basic capital-labour relation. We then attempt to show how this general conception of production is interwoven with the specific contingencies encountered by the semiconductor industry in its historical and geographical trajectory in a way that produces a unique, and highly problematical set of locational outcomes. Only in the light of a clarification of this issue is it possible to proceed effectively with the tasks of theoretical and empirical enquiry that now lie before us.

We pursue our broad agenda by describing the internal evolutionary dynamics of the semiconductor industry. We begin with a discussion of the central necessary internal relations of the industry (i.e. labour processes and technical change, cf. Marx 1967, Chapters 13, 14, 15), and we proceed by examining the industry's evolution in relation to the various national and international circumstances that have attended its overall development.

Labour Processes and Technical Change

The earliest growth of the solid-state semiconductor industry occurred in the years immediately following World War II. As the industry emerged technologically from vacuum tube production, the initial site of this growth was the North-East of the United States. It was here that the transistor was invented at AT&T's Bell Laboratories in 1947, and first commercially manufactured by another AT&T subsidiary, Western Electric in 1951 (Braun and Macdonald 1982). Throughout the 1950s more than half of all semiconductor production in the United States was concentrated in the New England and Middle Atlantic states (see Table 3.1.) A variety of factors, however, combined to discourage the continued concentration of the industry in the North-East and to stimulate its rapid development in Santa Clara County, California (with a secondary centre in Texas). We shall discuss these factors further below, but for the present, we are concerned to examine technical change in the industry in relation to labour processes and the structure of the labour force.

The growth of the semiconductor industry in Santa Clara County from the late 1950s coincided with the almost total displacement of the vacuum tube by transistors and diodes in both military and commercial markets. This technological shift not only had enormous implications for the size, cost and

Table 3.1 Geographical Origin of U.S. Semiconductor Shipments 1958-1972

Region	1958	1964	1967	1972
United States, total	100%	100%	100%	100%
New England	23	17		17
			43	
Middle Atlantic	36	31		21
New Jersey	10	4		7
New York	13	10		1
Pennsylvania	13	17		13
North Central				3
	28	25	22	
South				23
West	13	27	35	37
California				23
Other				14

Source: United States Department of Commerce (1979, p.7)

reliability of various forms of electronic hardware, but also for the type of labour processes associated with production. The different labour processes within semiconductor production were relatively technically disarticulated, (i.e. separated from one another as relatively discrete clusters of work tasks) and hence, potentially, spatially separable. These component labour processes consisted of (a) research and development (which in any case had been separated from other labour processes relatively early in the development of most branches of industrial production cf. Braverman 1974, Gorz 1976), (b) mask making, (c) wafer fabrication, (d) assembly and (e) final testing.

The development of these component labour processes created a polarisation in the skill requirements of the industry. R&D is strictly a preserve of highly qualified and creative scientists and engineers. Wafer fabrication too (especially the masking process) calls for highly skilled labour, though recently it has begun to make use of increasing numbers of semi-skilled workers. Assembly and final testing are given over almost entirely to unskilled labour (Keller 1981), though some skilled and semi-skilled supervisory labour is also employed in these processes.

This marked polarisation in the skill structure of the labour force was compounded by subsequent technological innovations, which revolutionised the whole organisational structure of the industry during the 1960s. Of particular importance have been the introduction of (a) the planar process for wafer fabrication (first developed by Fairchild Semiconductor in 1958). Because it facilitates the superimposition of one circuit over another on the wafer, this process was a necessary precondition for the successful commercial development of (b) the integrated circuit (first produced in quantity by Fairchild in 1962-63). By the end of the 1960s the integrated circuit had substantially replaced the transistor as the basic building block of electronic machines. It accordingly became the staple output on which the spectacular success of Santa Clara's semiconductor industry was based. Subsequent developments (in the 1970s) such as (c) the production of microprocessors and (d) the move to large scale and then very large scale integrated circuitry have had additional implications for the ratios of technical to unskilled labour power in the industry. The net result of these developments, has been an intensification of labour market segmentation in the industry as a whole. Hence, the proportion of professional/technical workers in the industry has tended to increase over time, while the proportion of semiskilled and unskilled workers has tended to remain fairly stable. By contrast, the phenomenon of a 'disappearing middle' of skilled production workers is increasingly evident. Some of these trends are highlighted in Table 3.2 which shows the occupational structure of the electronic component and equipment industry in Santa Clara County (of which semiconductors form a significant part) over the 1950s and 1960s.

This polarisation of the skill structure of the industry has led over the last couple of decades to the emergence of socially and spatially segregated labour markets. These segregated labour markets are also marked by particular gender and racial structures. Specifically, white males have tended to fill vacancies for scientists, engineers and technicians, while non-white females have predominated in the market for unskilled production labour. At the same time, the structure of the industry's labour market has had implications for the patterns of urban development and for the contradictions associated with it. These processes have had their most obvious impact in the industry's core complex, Santa Clara County, but to a greater or lesser extent have been played out across the globe subsequent to the development of offshore production.

Table 3.2 Occupational Structure of the Electronic Component and Equipment Industry in Santa Clara County, 1950–1970

Occupational Group and Gender of Employees	Percentage of Total Employment		
	1950	1960	1970
Non-Production Workers:	38.5	55.8	59.2
Professional/Technical	19.4	34.5	34.7
women	0.9	2.8	4.0
men	18.5	31.7	30.7
Managerial/Sales	4.5	7.0	8.7
women	-	0.5	0.5
men	4.5	6.5	8.2
Clerical	14.6	14.3	15.8
women	8.7	9.8	11.9
Production Workers:	61.5	44.2	40.8
Skilled/Supervisors	27.1	17.8	11.5
women	0.7	1.0	1.1
men	26.4	16.8	10.4
Semi-skilled/Unskilled	34.3	26.4	29.3
women	8.3	14.9	19.2
men	26.0	11.5	10.1

Source: Keller (1981, p.95)

We have indicated some of the features of the labour processes associated with semiconductor production. We need to add that a number of scholars, building on Marx's insights on the significance of the gradual replacement of the formal by the real subsumption of labour by capital (cf. Aronowitz 1978), have argued that capitalist industrial development necessarily involves regular attempts to reorganise and streamline the labour process. These attempts seek both to lower wage costs and increase productivity, and at the same time to eliminate, or at least neutralise, oppositional elements in the workforce and strengthen managerial control over production (see amongst others, Tronti 1972, Braverman 1974, Friedman 1977, Edwards 1979, Henderson and Cohen 1979, Nichols 1980, Thompson 1983). Attempts to reconstruct labour processes have typically involved organisational changes such as Taylorism and Fordism and/or the application of new technologies. Although both these forms of managerial intervention have been present in the semiconductor industry, as Keller (1981) has shown, the latter (for instance in the form of computer controlled automatic bonding machines) has tended thus far to dominate. This is scarcely surprising in an industry as dependent for its success on technological innovation as the semiconductor industry has been. These developments have also had important implications for the spatial development of the industry. It is to these questions, initially in terms the emergence of the industry's primary location - 'Silicon Valley' - that we now turn.

The Formation of the Core Complex: The Silicon Valley Phenomenon

It is often the case that the locational development of an industry is broadly divisible into two major phases namely, a first phase characterised by spatial convergence and concentration of producers, and a second phase characterised by decentralisation (cf. Hoover 1937, Norton and Rees 1979, Scott 1980, 1982). The American semiconductor industry is no exception to this general tendency. In the present section, we deal with the dynamics and manifestations of the first phase of development in the semiconductor industry; this involves above all an analysis of the steady build-up of semiconductor plants and ancillary services in the Santa Clara Valley of California from the mid-1950s to the early 1970s. In subsequent sections of this paper, we consider the industry's second phase which has been marked by a massive decentralisation of production facilities on both a national and a world scale.

The Emergence of the Semiconductor Industry in California.

In the previous section we mentioned that the growth of the semiconductor industry coincided with a locational shift from the North-Eastern states to Santa Clara County. There were a number of reasons for this shift.

In the first place, most of the large electrical and electronic apparatus producers in the North-East were already by the early 1950s strongly unionised (Troutman 1980, Bluestone and Harrison 1982). There was little to prevent the eventual incursion of union organisation into the semiconductor sector had it finally settled in the North-East. In the second place, the entire U.S. aerospace-defence industry

49

was by the late 1950s growing with great rapidity in the South-West of the country, and it started to draw into its geographical orbit a surrounding constellation of high technology industries, including the semiconductor industry.

Thus, when William Shockley (one of the inventors of the transistor) established the first semiconductor plant in Santa Clara County, just south of San Francisco in 1955, he alighted on a location where the prospects for the new industry were especially bright. Among other advantages of the location were the large teaching and research centres of Stanford University and the University of California which provided trained scientists and engineers, essential to the more technologically advanced sectors of semiconductor production. A note of methodological caution is necessary at once, however. Much of the literature on the development of the semiconductor industry in Santa Clara County puts great explanatory emphasis on Shockley's locational decision and on the proximity of major universities (cf. Bernstein et al 1977, Braun and Macdonald 1982, Saxenian 1983a, 1983b). We must be careful in ascribing causal powers to these phenomena. Shockley certainly planted the seed of the future development of what came to be known as Silicon Valley, but there can in general be no guarantee that any initial decision of this sort will come to social and economic fruition. Such fruition depends upon a series of additional necessary conditions of development. At the same time, the major educational establishments of the area were not at the outset already fully geared up to produce the research and manpower needs of the new industry; on the contrary, they were at best only potentially capable of meeting these needs as and when they concretely appeared, and the local universities and research centres tended to develop pari passu along with the industry as a whole. The point here is that wherever the industry may finally have settled down, we would expect to find local educational establishments adjusting their curricula and programmes accordingly (cf. Noble 1977 for the role of universities in an earlier period of industrial development).

If Shockley's initial decision and the proximity of major universities to Santa Clara County are really only contingencies, what then constitutes the internal necessary conditions for the consolidation of the industry in central California? We suggest that the development of Silicon Valley can most effectively be comprehended as a process of the

reproduction, growth, and transformation of a
localised territorial complex of productive labour
and social activity. By this we mean that there are
determinate dynamics of capitalist industrial and
urban development which ensure that any industry
whose market is growing with great rapidity will tend
to become rooted ever more insistently in a limited
number of areas. There are two major reasons for
this. One is that rapid horizontal and vertical
disintegration of production (especially during the
early stages of growth) lead to marked agglomeration
economies and Verdoorn effects in specific regions.
The other is that social and spatial reproduction of
the labour force is facilitated by the process of
geographical concentration. As these dynamics
proceed, an initial locational decision that may well
have been nothing more than a caprice, may turn out
to be the nucleus of a steadily self-confirming focus
of specialised production, (cf. Scott 1983). We now
need to look at these dynamics in considerably
greater detail.

**The Development of the Silicon Valley Production
Complex.** The first thing to note about the initial
development of Silicon Valley, from the late 1950s to
the early 1970s, is the extraordinary rate of
horizontal disintegration of the semiconductor
industry. Presumably this horizontal disintegration
occurred as a result of limited internal economies of
scale in individual production units in the context
of an overall market that was expanding at an average
rate of over 15% per annum in real terms (see Table
3.3). The outward form of this horizontal
disintegration was a series of "spin-offs" from pre-
existing firms. Thus, as Saxenian (1981) points out,
Fairchild Semiconductor was the progenitor of no
fewer than 50 companies in the Silicon Valley area
that were spun off in the two decades between 1959
and 1979. These proliferating companies constituted
the core of the developing production complex of
Silicon Valley. They were locked deeply into
competition with one another and they engaged in
advanced forms of product differentiation.
    As all of this was occurring, vertical
disintegration was also proceeding apace. In
conformity with Adam Smith's celebrated dictum (with
its specific analytical shape as defined by Stigler
1951), that the division of labour is limited by the
extent of the market, the rapidly rising demand for
semiconductors led steadily to a marked breakdown of

Table 3.3: U.S. Domestic Semiconductor Shipments by
Major Product Class, 1960-1970, (millions of current
dollars)

Year	Diodes rectifiers and related devices	Transistors	Integrated Circuits	Total U.S. domestic Shipments
1960	228	314	29	571
1961	249	316	38	603
1962	268	303	67	638
1963	282	312	190	784
1964	312	323	288	923
1965	379	426	317	1122
1966	471	504	492	1467
1967	444	434	505	1383
1968	420	427	568	1415
1969	464	472	751	1687
1970	421	411	888	1720

Source: United States Department of Commerce (1979,
p.39)

production activities into many specialised units.
Thus a wide variety of input services and
subcontracting activities now made their definite
appearance within the Silicon Valley complex. A
recent business directory indicates that the
following direct inputs (among others) are all
currently being produced in the Silicon Valley area:
automatic production machinery, testing equipment,
measuring devices, high vacuum equipment, encapsul-
ation materials, bonding materials, silicon crystals
and wafers, photomasks, epitaxial systems, metal
plating, deposition and etching services, circuit
designs, R&D services, and so on. This marked
vertical disintegration of the entire production
complex served at once to reduce the risks associated
with capital investment in a highly competitive
market, and (by externalising and pooling many
individual demands) to lower the fixed costs that
producers have to face.

The overall result of this marked horizontal and
vertical breakdown of production activities was the
development over the 1950s and 1960s of an intricate
system of specialised and ever-changing transactions
between producing units. Producers were (and are
still today) locked into a labyrinth of costly
materials flows and face-to-face contacts, and the
sheer weight of the spatial costs of these

interactions has encouraged plants of all kinds to converge towards one another around their own centre of gravity. The pressures on producers to cluster geographically were all the more irresistible in the earlier period of development of Silicon Valley when production technologies were extremely unstable and susceptible to brusque changes with correspondingly rapid variations in linkage structures. In these ways, the complex grew and local industrial land use became increasingly more dense. The industry's highly localised character has been underpinned both by the development of specialised banking services and venture capital firms, and by the many downstream consumers of semiconductors in the San Francisco Bay Area (communications firms, electronic instruments producers, aerospace, computer manufacturers etc.). Further consolidation of the complex was secured by processes of local reproduction of particular forms of labour power.

Social and Spatial Reproduction of the Labour Force.

In the previous section, we suggested that the development of the semiconductor industry's component labour processes resulted in a polarisation in the skill structure of the labour force and subsequently the emergence of socially and spatially segregated labour markets. Nowhere was this more the case than in Santa Clara County. As the industry grew over the 1960s and 1970s and the demands for its products increased, so the total demand for labour also grew. On the one hand, the industry's demands for unskilled and semiskilled labour was largely filled by immigrant female Latino and Asian (especially Filipino) workers (Tables 3.4 and 3.5) who resided in the San Jose area of the County, some distance from the centre of production in such north-County cities as Palo Alto, Mountain View and Sunnyvale. On the other hand, the industry's demands for highly trained scientists, engineers and technicians have been filled largely by white male graduates of local universities and colleges (Table 3.5), who tend to reside in relatively close proximity to the semiconductor plants and laboratories (Saxenian 1981).

One of the consequences of the structure and social composition of the semiconductor labour force, has been the relative absence of trade union organisation. Of American semiconductor workers who are unionised, 96% are employed in the North-East of the country by large, integrated multinationals such

Table 3.4: Trends in Electronics Employment in the San Jose Standard Metropolitan Statistical Area*, 1966-78

Year	A All Employees	B Blue Collar Employees	C Women in Blue Collar Work	D Racial Minorities in Blue Collar Work	B as % of A	C as % of B	D as % of B
1966	15,317	7,495	5,272	1,731	48.9	70.3	23.1
1969	19,758	6,232	3,701	2,516	31.5	59.4	24.3
1970	25,610	7,960	5,498	2,533	31.1	69.1	31.8
1971	24,504	7,493	5,216	2,142	30.6	69.6	28.6
1972	24,601	8,015	5,925	2,487	32.6	73.9	31.0
1973	33,420	12,813	9,640	4,326	38.3	75.2	33.8
1974	38,122	16,175	12,488	5,523	42.4	77.2	34.1
1975	39,852	13,072	9,520	4,552	32.8	72.8	34.8
1978	41,088	14,391	10,221	6,444	35.0	71.0	44.8

* Employment principally in Silicon Valley

Source: Snow (1980 pp. 43-45)

Table 3.5: Santa Clara County High Technology Employment* by Race and Sex, 1980

Occupational Category	Size (Per Cent)	Sex (Per Cent) Male	Female	White	Black	Race (Per Cent) Hispanic	Asian	Am. Indian
Managers	14	85	15	88	2	4	5	1
Professionals	20	82	19	83	2	3	12	1
Technicians	15	75	25	71	3	10	15	1
Sales	2	67	33	91	2	3	3	1
Clerical	15	19	81	77	6	10	6	1
Craft	7	56	44	63	6	17	14	1
Operatives	24	31	69	49	9	23	19	1
Labourers	2	38	61	41	8	34	17	1
Services	1	86	14	49	12	26	13	1
Totals	100	57	43	70	5	12	12	1

* Predominantly in Electronics

Source: Siegel & Borock (1982, Table 9 p. 46)

as AT&T and ITT. As of early 1980, none of the workers employed by the 'merchant' semiconductor producers in Silicon Valley were unionised (Troutman 1980). Partly as a result of the lack of unionisation, wage rates for production workers in the Valley were in 1977 between 31% and 61% lower than for their unionised colleagues in the North-East (Troutman 1980).

During the 1970s the semiconductor labour markets have, if anything, become increasingly socially and spatially segregated. The production workforce has continued to be reproduced by drawing in workers of Third World origin, a growing proportion of whom are now illegal immigrants (Snow 1980, Katz and Kemnitzer 1982). While the social reproduction of this part of the labour force has not thus far produced significant problems for the semiconductor companies, the spatial dynamics of their reproduction has. Specifically these workers have been subject to rising costs in terms of time and money as a result of having to commute from the relatively distant, but low cost housing areas of San Jose and adjacent communities. It is these problems of commutation rather than the routinised work, or poor wages per se, which seem to have led to shortages of unskilled labour, and high turnover rates in the Valley's semiconductor labour force (Saxenian 1981).

The main bottleneck on the further development of the industry in Silicon Valley in fact has been the reproduction and supply of highly skilled engineering and technical labour. As indicated above, Stanford University in particular accommodated itself from the start to the basic manpower needs of the industry. Stanford has produced a constant (but always numerically inadequate) stream of graduates at every level of academic attainment from B.Sc. to Ph.D. who have moved directly into jobs in the semiconductor industry. As we have suggested, this phenomenon can scarcely be seen as an 'independent variable' but rather as a subjacent moment of the entire developmental process of Silicon Valley in general. There is a widespread tendency for local colleges and universities to adapt their educational programmes to the peculiarities of local industrial activity, and Silicon Valley has been no exception in this regard; moreover, by acting in this way, they help to socialise the costs of specialised manpower training thereby lowering manufacturers' overheads. The skills of the labour force are finely honed in the work-place, and a many-sided local labour market comes into being alongside the dominant production complex. Firms then are able to fill vacancies as they arise from within this pooled labour resource. At the same time, the labour force must be housed and appropriate forms of communal and recreational activity allowed to develop. It is especially important for the continued viability of the complex to ensure that the delicate norms of neighbourhood activity and environmental quality necessary for the social reproduction of the upper echelons of the labour force be secured. This imperative has been all the more pressing in Silicon Valley in view of the circumstances that upwards of 40% of all workers in the industry has typically consisted of highly educated and highly paid white collar workers. Saxenian (1981) has described the communities inhabited by these workers as typified by low-density suburban tracts with expensive housing and a wide variety of recreational opportunities. This kind of development is especially characteristic of the affluent northern and western foothill cities of Santa Clara County where a spacious and semi-rural ambience prevails. Local municipalities and planning agencies have helped to underpin this state of affairs by imposing appropriate zoning provisions and density controls. In these ways, then, the form and substance of local urbanisation processes have also helped to cement the complex into a functioning, viable whole.

None of these processes of growth, development, and reproduction proceeds unproblematically, however. Sooner or later various limits to further expansion of the complex begin to make their appearance.

The Predicaments of Concentrated Territorial Development. We have seen that industrial complexes tend to grow, in the first instance, at least, via the dynamics of horizontal and vertical disintegration of labour processes and via appropriate social and spatial reproduction of their associated labour forces. These dynamics tend to produce a situation in which costs of production on the terrain of any given complex fall steadily in relation to all other possible locations, and this then propels the complex forward into yet more advanced stages of development. As this occurs, the complex begins increasingly to encounter internally-generated barriers to its own further growth, and problems mount as more and more productive activities and population pile up in the local area. In Silicon Valley, these problems started to become especially pronounced after the late 1960s: shortages of engineering and technical labour became chronic, wages of all sections of the labour force escalated upwards, land values increased, and a severe shortfall of adequate housing seemed to become endemic to the whole area (Bernstein et al 1977, Saxenian 1981). In view of these difficulties, it was not long before representatives of the semiconductor industry were seeking out ways of emancipating themselves from that which they had created at the outset, that is, (by means of technical and organisational restructuring) of dispersing new investment in particular component labour processes to profitable production locations elsewhere.

Before we continue with our account of the spatial development of the industry, we need to return to our examination of technical and social processes underlying this development.

Capital Deepening, Markets, and the Determinants of Internationalisation.

From its very inception, the semiconductor industry has tended to become increasingly capital intensive. The cost of setting up a state-of-the-art semiconductor manufacturing plant in 1965 was only about $1 million; by 1980, this cost had escalated to $50 million (Saxenian 1981). Simultaneously, the character of the industry was changing. This was reflected in part in the growing proportion of integrated circuits (as opposed to discrete devices) in total semiconductor output, as it was (in part) in the increasing tendency to standardisation of outputs in important market segments.

In the early years of the industry, many relatively small firms emerged in Santa Clara County around an insistent process of horizontal and vertical disintegration in which, in addition, firms provided more or less customised products in response to specific market 'niches'. As we have already demonstrated, the result of this process was not only a mushrooming of producers in Santa Clara County, but also the development of an industrial system based on small-batch production. In this kind of an economic environment firms were limited in the levels of automation that they could achieve. In particular, it was important for them to be able to switch easily from one product line to another, and this necessitated relatively low capital-labour ratios

(Keller 1981). Further, in the early years, the number of bonding operations to be carried out per device in the assembly stage was small, and so manual labour was not yet the barrier to profitability and expansion that it was later to become. In fact, this barrier only really became decisive in the late 1970s, and automated assembly procedures began to be assimilated on a large scale into production activities in order to counteract it.

Even so, from the early 1960s, tendencies to technical change, capital deepening, and the standardisation of product lines were apparent. This was in part associated with a relative decline in military purchases of semiconductors and a relative growth in civilian markets where product demands were less specialised. In 1960, military end-users still consumed 50% of US semiconductor production, but by 1966, this had dropped to 30%, and then by 1972 to 24% (Braun and Macdonald 1982). This relative decline in military markets had three major corollaries. First, as noted, it was associated with an increased relative (and absolute) level of large-scale, standardised demands. Second, it induced stronger price-competition and the necessity for more stringent cost-cutting measures (since reliability more than cost is the concern of military purchasers). Third, it opened the way to the internationalisation of production, for military procurements, by federal law, are almost entirely restricted to domestic manufacture (Snow 1980). All of this has encouraged a series of significant locational shifts in the industry.

Determinants of the Internationalisation of the American Semiconductor Industry. We have already indicated that the production of semiconductors involves five technically disarticulated labour processes, i.e. R&D, mask making, wafer fabrication, assembly, and testing. The different labour processes in the industry have widely varying needs in terms of capital investment, labour skills, specialised inputs, and so on. Although these component labour processes were in general all found in geographical association with one another in the early years of the industry's growth in Santa Clara County, the fact that they were technically disarticulated meant that they could be organisat-ionally and hence spatially separated. To be sure, there are definite internal economies of scope and scale that tend to keep all the labour processes in the semiconductor industry unified within the single firm. Nevertheless, if linkage costs on their transactional activities with one another are sufficiently low, then the possibility of their spatial dispersal can be made real. Cheap air transport and telecommunications have potentiated exactly this outcome.

That said, there are other strong determinants of the location process in the semiconductor industry. The precise ways in which these determinants play themselves out depend very much on which type of labour process we are talking about. R&D requires ready access to highly trained scientists and engineers for its successful operation; as a corollary, it needs the kind of local urban/environmental conditions (such as are found in and around Silicon Valley) which can sustain the effective social reproduction of this form of labour; and, of course, the localised training of such labour in universities and research institutes is a further major asset. Many of the same conditions apply to the location of mask making and wafer fabrication facilities. However, there has also been a moderate internationalisation of these facilities, particularly wafer fabrication, in part in order to evade certain kinds of tariff barriers and penetrate increasingly lucrative markets. Thus companies such as Motorola, National Semiconductor, Hughes and General Instrument have located wafer fabrication plants in Britain as a means of successfully penetrating the high (17%) E.E.C. tariff barrier on semiconductors and of therefore tapping the European market. Assembly and testing are typically unskilled operations and these have been increasingly relocated out to the periphery of the world system over the last two decades.

All the component labour processes of the industry were effectively confined to the United States in the early period of development when markets were limited and specialised, and the military constituted the principal end-user. With the insistent rise of more standardised commercial markets, the conditions for internationalisation of parts of the industry were brought to fruition. First assembly, then testing functions dispersed to other parts of the globe, especially to Latin America, parts of Europe, and above all to South East Asia. Then, in response to the E.E.C. tariff barrier and a growing market, some wafer fabrication activities followed on to Europe. Given the large amounts of labour employed in assembly and to some extent testing, the low wages (given potentially high labour productivity) at the sites selected in the periphery, were undoubtedly a prime locational determinant (cf. Troutman 1980, Frobel et al 1980, Rada 1982, Ernst 1983). We must add to this the remark that political stability and the security of capital investments have become increasingly important in those offshore locations where firms have sought technologically to upgrade their plants.

Finally, we should observe that American corporations chose to exploit peripheral locations not by means of licensing activities or subcontracting arrangements, but predominantly by means of direct investment. Only they themselves possessed the organisational and managerial capacity for effective profitable implementation of the necessary labour tasks. They therefore needed to keep full proprietary control over offshore production units in order to secure the full advantages to be reaped from internationalisation (cf. Dunning's 1981 theory of the multinational corporation).

It is only through an investigation of the articulation of these various determinants as they have changed over time and impacted on the industry's component labour processes, that we can comprehend the waves of internationalisation indicated above and the associated emergent international division of labour in the semiconductor industry. We now consider in some detail the specifics of the offshore development of the industry. In the next two sections of the paper, we deal with the special cases of the development of U.S.-owned branch plants in South East Asia and Scotland.

Internationalisation and Spatial Division of Labour: The Case of South-East Asia

Observe at the outset that the first intimations of the internationalisation of the industry appeared when Fairchild Semiconductor set up a plant to assemble discrete devices in Hong Kong in 1962 (Siegel and Grossman 1978). By the end of the 1970s there were semiconductor factories scattered across selected locations in the Third World, but especially in Latin America and South-East Asia. In addition to Hong Kong, U.S. semiconductor branch plants were to be found in Singapore, Malaysia, Taiwan, South Korea, Thailand, the Philippines, and Indonesia. At the same time, a number of U.S. companies established production facilities in the peripheral regions of Europe, with Scotland being the preferred recipient

(Siegel 1980).

These locational trends have been played out in different ways in different areas, depending on many varying local and contingent circumstances. In what now follows, we attempt to deepen our understanding of the process of internationalisation by looking in some detail at the South-East Asian case.

The Development of the Semiconductor Industry in South-East Asia. We begin by attempting to describe the circumstances that led to the initial establishment of the industry in Hong Kong. We then show how it diffused out from Hong Kong and how this diffusion has brought in its train a definite subregional division of labour.

We earlier discussed the internal production conditions that tend to encourage dispersal of component labour processes in the semiconductor industry. But these conditions are also intertwined with a series of contingent circumstances that lead to definite, realised locational outcomes. In the case of South-East Asia the most important of these circumstances by far has been the presence of enormous pools of cheap and underemployed labour (Rada 1982). But the mere existence of cheap labour was in itself not a sufficient locational inducement. Hong Kong had a number of additional special advantages, which made it a particularly attractive location. (Similar advantages in Singapore meant that it was also to develop a flourishing semiconductor industry at an early stage). These advantages included (a) political stability, (b) an open financial system with no limits on the repatriation of profits, and (c) excellent telecommunications and air transport facilities (Henderson and Cohen 1982b). In addition, Hong Kong had a further crucial advantage: Over the 1950s, it had developed a flourishing industrial economy based on textiles, garments, plastics, and other labour-intensive forms of production. By the late 1950s, Hong Kong had also become a major location for U.S.-owned radio assembly and production of the cheaper varieties of consumer electronics (Chen 1971). This meant that Hong Kong possessed by the early 1960s a work force that was habituated to the kinds of labour processes characteristic of semiconductor assembly, as opposed to labour processes like, say, iron and steel or shipbuilding that formed the basis of early Korean industrialisation, (cf. Henderson and Cohen 1982a for a general account of the significance of labour habituation). Furthermore, the existence of a flourishing informal sector in Hong Kong at that time helped to keep industrial wages down to a level that was quite comparable to wages in many other Third World areas. Finally, Hong Kong was also able to supply the small but crucial demand for qualified engineers and technicians necessary for the successful operation of semiconductor assembly functions.

Fairchild and the firms that followed it to South-East Asia established a standard pattern of internationalisation in the industry. They employed for the most part young unmarried female workers (Lim 1978, Grossman 1979). They imported wafers fabricated in Silicon Valley and air freighted to South-East Asia. They then assembled the discrete devices or integrated circuits obtained from these wafers. And then they air freighted the assembled units back to Silicon Valley for final testing and marketing. The emergence of this characteristic division of labour was reinforced by U.S. tariff regulations 806.30 and 807.00 which relieved U.S. (re-)importers of semiconductors of all duty except on the foreign value added.

From its beginnings in 1962 until the mid 1970s, semiconductor production in South-East Asia took place in wholly-owned U.S. branch plants. The reasons for firms' preference for this form of production (in contrast, say, to joint ventures or licensing operations) seem clear enough. Reliable production methods and high yields (in circumstances where both production technologies and product specifications were changing rapidly) depended on rigorous technical and managerial control. Even at the present time, offshore production of semiconductors is overwhelmingly conducted in wholly-owned branch plants. There has, however, been a subsidiary development of subcontracting relations within the South-East Asian system (particularly in the Philippines). Plants in places like Hong Kong and Singapore have begun to subcontract out assembly of the less sophisticated types of semiconductor to independent producers, especially at times when the former have been unable to cope themselves with excess demand (Paglaban 1978).

The Sub-Regional Division of Labour in the South-East Asian Semiconductor Industry. Throughout the 1960s and up to the 1970s, the pattern of the international division of labour described above was repeated many

times over in South-East Asia. In each case the basic
form of U.S. penetration of the local economic system
was the same. The offshore plants specialised in
intermediate (assembly) processes, but basic
managerial and technical control as well as R&D, mask
making and wafer fabrication, remained firmly
implanted in the United States, as did much of the
product finishing process, until relatively recent
years, at least. Nevertheless, an internal
subregional division of labour within South-East
Asia did start to come about. In the more advanced
centres of production - specifically, Hong Kong and
Singapore - branch plants began to specialise more
and more in the assembly of relatively small-batch
high cost semiconductors. After about 1971 and the
first incursions (by National Semiconductor) of U.S.
semiconductor branch plants into Malaysia, the less
developed locations (Indonesia, Thailand, the
Philippines, in addition to Malaysia) took over much
of the large batch, standardised, low quality
production activities (Siegel 1979, 1980). At the
same time, Hong Kong and Singapore became important
centres of final testing, not only for their own
products, but also for those made elsewhere in South-
East Asia. Testing of course is a delicate operation
that calls for intense levels of managerial and
technical supervision, and only the main centres of
production could meet the necessary quality control
standards. Motorola has been especially active in
developing this aspect of the subregional division of
labour and since the early 1970s its Hong Kong plant
has been the main local testing centre for the
company's output from its Philippine, Malaysian, and
South Korean plants. Subsequently, Fairchild and
National Semiconductor have also developed major
testing facilities in Hong Kong and Singapore. We
should note that our field research has indicated to
us that in very recent years there has been some
outward diffusion of low-level testing from Hong Kong
and Singapore. National Semiconductor, for instance,
now tests as well as assembles integrated circuits
(made from wafers fabricated at its Scottish plant)
at its plants in Malaysia and Thailand. In spite of
this, however, it is still Hong Kong and Singapore
that remain the prime South-East Asian foci of this
particular activity.

Why, we may ask, has this peculiar subregional
division of labour occurred? Why, in other words, has
more and more of the investment in assembly plants
for large-batch standardised low quality outputs
tended to go to an increasing degree to Indonesia,

Malaysia, the Philippines, and so on, while Hong Kong and Singapore have tended to be upgraded as to the quality and complexity of their production technologies and labour processes? There are two major reasons for this. First, throughout the 1970s, labour costs in Hong Kong and Singapore increased dramatically relative to other parts of the region. Thus as Table 3.6 shows, both Hong Kong's and Singapore's manufacturing wages shifted strongly in the direction of U.S. manufacturing wages between 1969 and 1980. By contrast, manufacturing wage rates in the Philippines, Indonesia, and Thailand (and, we might add, Malaysia) were still far below those of the United States in 1980, and, in fact, considerably below those of Hong Kong and Singapore (and South Korea and Taiwan also). Thus, in the continued insistent search for cheap labour (combined with reasonable levels of political security) much new assembly work has shifted to these low-wage countries. Second, Hong Kong and Singapore (along with South Korea and Taiwan) had for long been able to supply well-trained technicians, engineers, and scientists. This is a reflection in part of their more advanced level of development, as it is also in part of their advanced educational systems that can generate large numbers of highly qualified personnel. Thus, critical testing activities can be carried on in these centres with quite high levels of reliability. By the same token, Hong Kong and Singapore are also in the process of becoming local centres for the provision of customer services throughout the expanding South-East and East Asian market. This subregional division of labour can be seen in value-added data for semiconductor imports into the United States. In 1979 the value-added to semiconductors partially processed in Hong Kong and imported into the United States was 56% of their total price. For semiconductors imported from Singapore (and Taiwan), the corresponding figure was 46%. For Malaysia, Korea and the Philippines, the value added was, respectively, 45%, 39%, and 32% of price (Ernst 1983).

We must add to these remarks the comment that Hong Kong and Singapore are also becoming important centres of local sales and marketing in the industry. Today, the regional headquarters of Motorola is in Hong Kong and that of National Semiconductor in Singapore. Thus these cities function as localised centres of management and control within the overall global pattern of the industry.

Table 3.6: Ratio of U.S. to foreign manufacturing
wages for a sample of South-East Asian countries

	1969	1980
Hong Kong	10.3	2.6
Korea	10.2	3.0
Taiwan	18.2	3.8
Singapore	11.1	4.7
Philippines	n.a.	7.9
Indonesia	n.a.	8.1
Thailand	n.a.	11.9
India	n.a.	14.7

Source: Rada (1982, p. 186)

Multiplier Effects and Local Growth Impulses. For
much of its brief history in South-East Asia, the
semiconductor industry has had remarkably little
impact on local economic structures. With the
exception of minor purchases, linkages to the local
economy have been slight; and employment has rarely
been numerically significant relative to total
population. However, the growth and development of
the industry in Hong Kong and a few other places,
together with the spatial reintegration of assembly
and testing does now seem to be encouraging the
incipient emergence of semiconductor industry
complexes. These complexes include firms that supply
many kinds of ancillary materials and services to
U.S.-owned (and other) semiconductor plants. They
also include equipment manufacturers, the notable
example here being the American firm of Kulicke and
Soffa (the leading manufacturer of automatic
semiconductor bonding equipment) which now has a
branch plant in Hong Kong. In spite of the
circumstance that the semiconductor complexes of
Hong Kong and other areas are still in their infancy,
they have nonetheless helped to provide the
conditions under which locally-owned semiconductor
firms have managed to come into existence. There are
three such firms in Hong Kong (reputedly backed by
the People's Republic of China) and at least one each
in Taiwan and South Korea (Siegel 1980, Neff 1982).
These firms use wafer masks produced in Silicon
Valley, but have in-house wafer diffusion
capabilities. They produce low-grade standardised
outputs for local markets. Furthermore, during the
last two or three years, a number of U.S. firms have
set up design facilities in Asia in order to service

the product requirements of their Asian customers. Motorola, Siliconix and Teledyne have established design sections in Hong Kong as has National Semiconductor in Singapore.

With the possible exceptions of Taiwan and South Korea, U.S. semiconductor firms have not developed design facilities elsewhere in South-East Asia. It may be significant to note that by 1983 Motorola was seriously contemplating construction of a wafer fabrication facility in Hong Kong in order to service its other South-East Asian plants, though up to the present time, no such facility in fact, has appeared. We also found in our field enquiries that some U.S.-owned semiconductor plants in Hong Kong are now vertically integrating downstream, and are starting to produce electronic sub-assemblies using not only locally-manufactured semiconductors, but also other components including printed circuit boards, switches, rectifiers, and so on. Thus the Hong Kong semiconductor complex is also beginning to merge into the larger and very definite local electronics complex. Similar developments also appear to be underway in Singapore (Lim and Pang 1982).

Internationalisation and the Spatial Division of Labour: The Case of Scotland

The European Market and Semiconductor Production System in General. In the last two decades, Western Europe has emerged as a major and growing market for microelectronic devices. This market is protected by restrictive E.E.C. trade barriers which impose a 17% duty on all imported semiconductors. Thus, in order to participate to the maximum possible extent in this market, many U.S. semiconductor firms have established branch plants at various European locations. Fairchild, for instance, set up a plant in West Germany, Motorola set one up in France, Texas Instruments in Germany, Italy and France, ITT in France and Germany, Siliconix in Wales, and so on. However the principal regions in Europe receiving U.S. semiconductor investment have been Ireland (Mostek, Analog Devices, General Electric and most recently AMD and Zilog), and above all (in terms of the value of investment) Scotland (Motorola, National Semiconductor, General Instrument, Hughes and most recently Burr-Brown) (Siegel 1980, Locate in Scotland 1983).

By and large, the patterns of U.S. direct investment in the European and Scottish semiconductor industry differs quite markedly from the South-East Asian case. In Europe, branch plants tend to specialise in wafer fabrication together with some customer-related design facilities. Recall that wafer fabrication is a relatively advanced element of the production process, and it requires very high levels of capital investment, highly skilled labour and rigorous quality control procedures. In Asia, as we saw, plants specialise in the less capital intensive and less demanding (in terms of skill requirements) functions of assembly and testing. In point of fact, wafers manufactured in Europe are commonly shipped to South-East Asia for assembly and testing and are then re-imported (incurring import duties only on the value added overseas) back into Europe. Two exceptions to this pattern are Siliconix which has a facility devoted exclusively to assembly and testing in South Wales, and Motorola whose Scottish plant is becoming relatively integrated with wafer fabrication, assembly, and testing functions. Note, however, that Siliconix is a specialised, low-volume, high value producer, and Motorola's assembly functions in Scotland are fully automated. By shipping out wafers for assembly in South-East Asia, plants save greatly on labour costs (which, for example, are about 60% lower in Hong Kong than in Scotland). Note that R&D still tends to remain heavily concentrated in the U.S., though some R&D functions (largely of a product development and application nature) have also been transplanted to Europe as part of the necessary infrastructure of customer service.

Semiconductor Production in Scotland. Given the development of a branch-plant semiconductor industry in Europe, why did Scotland emerge as the preferred location? There seem to be three major reasons underlying this outcome (cf. Henderson 1986).

First, in view of the circumstance that the basic function of the Scottish industry is wafer fabrication, the availability of high quality scientific and technical personnel is a precondition of successful operation. This, in turn, presupposes a local educational system that can continue to reproduce such personnel in significant numbers. As one of the primary foci of earlier rounds of British industrialisation, Scotland has typically possessed a large standing pool of scientists, engineers, and technicians. The local reproduction of these categories of labour, moreover, is assured by the deepening commitment of the Scottish Universities (Edinburgh, Glasgow and Herriot-Watt) in particular to high-technology research and training (Locate in Scotland 1983). This mirrors in many ways the analogous role of Stanford University in Silicon Valley. At the same time, engineers and technicians in Scotland command remarkably low wages; these are currently some 60% below the wage level for equivalent workers in Silicon Valley (Locate in Scotland 1983, Troutman 1980).

Second, much of Scotland's Central Valley has been officially designated a 'Special Development Area', and this has resulted in a number of financial advantages for the semiconductor industry. In order to attract new investment into central Scotland (as well as other deindustrialising and economically-depressed areas of Britain) the British Government provides up to 22% of initial development costs, as well as other incentives to new locators (Locate in Scotland 1983). Since the development of a wafer fabrication facility entails extremely heavy capital expenditure, this subsidy may have been a major attraction for the semiconductor industry. With the restructuring of the British Government's regional policy in late 1984, and the reduction in subsidies to much of the central belt of Scotland that this has produced, it remains to be seen what effect reduced subsidies will have on future semiconductor investment.

Third, just as we observed a self-reinforcing dynamic of territorial development in the Silicon Valley complex, so also do we observe the beginnings of a somewhat similar tendency in Central Scotland. In fact, the two locational determinants alluded to above do not in and of themselves constitute a unique attribute of central Scotland. Other parts of Britain, including Northern England, South Wales, and Northern Ireland have very similar locational conditions. In the case of Scotland, however, these conditions have been combined with the unique (and perhaps fortuitous) advantage of an early start, and the elementary outlines of an electronics production complex have appeared there before they have in the other areas. Rather like the Silicon Valley case there was also in central Scotland an existing corpus of plants with strong potential linkages to and from the semiconductor industry. This had begun with the British military market supplier, Ferranti, setting up a plant in 1943. It continued from the late 1940s with investment from American multinationals such as Burroughs, Honeywell, NCR, IBM and Hewlett Packard who developed plants in Scotland to manufacture electro-mechanical and electronic equipment for the European market (Firn and Roberts 1984).

Since the mid-1970s, the Central Valley of Scotland has developed increasingly as a high technology industrial complex with further rounds of investment from all the U.S. semiconductor firms established there as well as new semiconductor investment from a small Arizona firm, Burr-Brown and the Japanese giant, Nippon Electric. In addition, Scotland now possesses three indigenous wafer fabrication facilities, one of which (Integrated Power Systems) will soon begin to manufacture integrated circuits in commercial quantities (Firn and Roberts 1984, Hargrave 1985).

The growth of this complex has been underpinned by the development of local and foreign firms supplying a wide variety of components and subcontracting services to the industry. Most significant technologically has been the emergence of the Scottish mask-making firm, Compugraphics and the recent decision of the Japanese silicon wafer producer, Shin-Etsu Handotai, to set up a plant in the country (Baggott 1985).

Now, even though Scotland possesses many positive features attracting investment by U.S. semiconductor producers, it has one strong potentially negative aspect, and that is the heavily unionised and historically militant Scottish working class. The semiconductor houses, however, have successfully by-passed this potential problem by means of their highly selective employment policies.

With the exception of National Semiconductor which is situated in the declining shipbuilding centre, Greenock, all of the semiconductor houses have located their plants in the 'new' towns of East Kilbride (Motorola) Livingston (Burr-Brown, Nippon Electric) and Glenrothes (Hughes, General Instrument). In each case these towns constitute relatively isolated labour markets, and in particular are well removed from traditionally militant and socialist Glasgow area.

In addition to the highly trained scientists and engineers that they employ, the semiconductor plants also require large numbers of semiskilled and unskilled manual workers, for wafer fabrication is quite deskilled and labour-intensive in certain of its aspects. This manual labour is not drawn from the traditional (male) Scottish labour force, however, but from the female population, which employers tend to regard as more passive and malleable. In as far as the semiconductor plants are entirely non-union and have been practically strike free, there seems to be some evidence to support their contention. Persistently high levels of unemployment in Scotland over many years, however, must also have helped control labour militancy in the plants.

In as far as Scotland has been a recipient of more advanced kinds of branch plants than those that are typically implanted in South-East Asia, it is clearly in a more favourable (i.e. more dominant, more autonomous) position within the new international division of labour than the latter area. This is purely in relative terms, however. The Scottish industry itself remains not only subordinated to the United States in terms of organisational and managerial control, but also in terms of scientific and technological inputs. Thus, although U.S. firms have built wafer fabrication plants in Scotland, they have not as yet installed facilities to produce wafer masks. This is the most technologically sophisticated part of semiconductor manufacture, and the masks are still imported overwhelmingly from the United States. Moreover, wafers fabricated in Scotland are for the most part designed for the production of linear and memory function integrated circuits rather than for microprocessors, which constitute the most advanced kind of integrated circuit output. Wafers for microprocessors are still predominantly produced within the United States. Lastly, even though the Scottish plants have some customer-related design facilities and R&D functions, the most advanced forms of the latter continue to remain firmly tied to Silicon Valley and similar regions of the United States (Henderson 1986).

Retrospect and Prospect: The New Spatial Political Economy of Semiconductor Production

We have shown in the above account how a system of internal necessary relations (i.e. capitalist relations of production) in the semiconductor industry encountered and combined with a set of external contingent circumstances. The result has been the observable ever-varying history and geography of the industry as described (in part) above. We showed, in particular, how the industry concentrated at an early stage in its development in Silicon Valley; how it then began to disperse out from this primary locational focus to a series of peripheral locations; and how an international division of labour was established as a consequence. These processes have not and do not occur in the manner of a simple and smooth unfolding. On the contrary, they are marked by many discontinuities and rapid shifts of direction, and there is good reason to assume that this will continue to be the case in the future. We may ask, what impending or future developments are likely to bring about new changes in the overall spatial structure of the industry? Two major points need to be made.

First of all, we foresee for the immediate future a consolidation and intensification of the trends described above. This involves continued managerial and technical domination at the core along with continued diffusion for the more advanced forms of production to selected peripheral sites (like Hong Kong and Scotland) in response to growing markets in particular regions. At the same time, we expect to see a further diffusion down the spatial hierarchy of the most-routinised, least sophisticated (technologically simpler) forms of production to such Third World countries as the Philippines, Indonesia, and so on. We thus anticipate a consolidation of the current international division of labour in the semiconductor industry, namely, (a) a core-periphery articulation, and (b) a sub-regional articulation within the periphery such that we can think of there being sub-centres locally in control of a purely regional pattern of economic activity and interchange. In this process, we observe that at least some of these sub-centres (e.g. Hong Kong, Singapore, but much more importantly, Scotland) are beginning to develop as semi-autonomous production complexes, with definite linkages to local suppliers, subcontractors, and buyers. As the Asian and European markets continue to grow, this trend will certainly become more exaggerated. Within these complexes we also already observe indigenous entrepreneurs beginning to set up their own forms of semiconductor production. These critical interactions between many different kinds of economic agents in various countries suggest that any disturbances of the current political accommodations within the global semiconductor production system, may well have serious repercussions. This is all the more the case given the importance of microelectronics within contemporary capitalist production activities.

Secondly, both the global semiconductor system and the international political accommodations that have made it possible may well, in any case, be seriously perturbed by ongoing processes of technical change in the industry. In the long run, these processes will certainly create a further series of radical transformations of the industry. Here we point to a couple of highly indicative cases. On the one hand, new production technologies are beginning to appear in response to the development of very large scale integrated circuits (i.e. VLSI technologies). One of the characteristics of these kinds of circuits is that the bonding tasks associated with their assembly are extremely complicated by reason of the large number of external connections that must be dealt with in the labour process. Accordingly, it would seem as though the assembly of such circuits will become very highly automated. On the other hand, we expect to see, in any case, a continuation of the process of rapid capital-deepening in all phases of the industry at large. This is likely to go on even in those sectors of the industry that currently employ cheap female Third World labour. The consequence of all of this, of course, is likely to be a considerable reduction of the proportion of labour costs in final prices, if not an absolute and dramatic shedding of labour in many cases. But even more importantly, all of this technological change may well be associated with a resurgence in the growth of American domestic semiconductor production (in all its phases) combined with a significant repatriation of American capital. Some of these processes are already discernible in the choice of intermediate locations (i.e. in the American semi-periphery between Silicon Valley and the Third World) of relatively routinised wafer fabrication: in Utah (National Semiconductor, Signetics), and Arizona (National Semiconductor), for example. And both Motorola and Fairchild have expressed their intention to repatriate their assembly functions back to the United States. (Electronic News, April 26, 1982; Global Electronics Information Newsletter, December 1983)

Although these developments may not lead to the wholesale exodus of American semiconductor production from South-East Asia, Europe, or Latin America, it is precisely the assembly of the more technically advanced devices, destined for American customers, that is perhaps most likely to be repatriated back to the United States. Paradoxically, if this were to occur, it would impact most adversely on those semiconductor and electronics complexes, as in Hong Kong and Singapore, that seem to be approaching a capacity for self-sustained production.

Whatever the eventual outcome of these particular developments two things remain clear. Firstly, there do appear to be emergent cores within sub-regional divisions of labour, and the relative local dominance of these cores seems to be encouraging a number of definite growth centre effects. Secondly, whatever the economic, technological and organisational strengths of these cores, they, like the other parts of the global American semiconductor industry, ultimately remain subordinated to decisions taken in Silicon Valley and similar locations.

All of this suggests that the form of the new international division of labour as described by Frobel et al (1980) is only an early stage of a rapidly evolving system. What we see in the case of the global semiconductor industry today is not just a set of fairly simple intra-firm transactions and inter-firm subcontracting relations across international boundaries but the development of a complicated social and locational system. This system may be described as a hierarchically-ordered and spatially-differentiated set of integrated production relations cemented by an underlying set of relations of political domination and subordination. The dynamics of this system are of the utmost interest and significance, and their practical working out is intensely problematical.

Any attempt to understand this phenomenon of necessity involves invocation of a wide variety of issues: the multinational corporation, location theory, industrial organisation, technical change, the labour process, worker resistance, urbanisation and regional development, and all the rest. Quite apart from its immediate social and political interest, the microelectronics industry (and other industries like it) poses innumerable research problems for the future.

Acknowledgements
This paper draws in part on Jeff Henderson's research on the socio-spatial dynamics of the changing international division of labour and specifically on his work on semiconductor production in California, South-East Asia and Scotland. This research has been supported by grants from the Lipmann Trust, London, the University of Hong Kong's Urban Studies and Urban Planning Trust Fund, and the Board of Management of Urban Studies. The paper also draws on Allen Scott's interests in industrial location and particularly on

his research project on the patterns of industrial location in the modern metropolis (funded by the National Science Foundation under grant number SES 8204376). The paper was originally drafted when Allen Scott was Visiting Research Fellow in the Centre of Urban Studies and Urban Planning, University of Hong Kong. Funds in support of his visit were provided by the Croucher Foundation. Final revisions to the paper were done when Jeff Henderson was University Visiting Fellow in the School of Environmental Planning, University of Melbourne. We are most grateful to these various agencies for their financial support. We are also grateful to Doreen Massey, Chris Pickvance, Jonathan Schiffer and Lenny Siegel for their helpful comments on earlier versions of the paper.

Abstract

The paper deals with the organisational and spatial development of the semiconductor industry in the United States, and with the subsequent diffusion of parts of the industry to offshore locations. Theoretical and methodological preliminaries to the task of analysing these issues are discussed. The early spatial concentration of the industry in Silicon Valley is investigated and it is shown how the industry constituted the core of a strongly-developed economic and social complex at this location. It is argued that the component labour processes within semiconductor production are relatively disarticulated in technical terms. As a result, in the context of the development of a number of economic and social determinants, there has been a major shift of specific labour processes out to various peripheral locations. The development of the industry in the particular cases of South-East Asia and Scotland is considered. The paper concludes with a brief statement on the relevance of the study to the emerging theory of the new international division of labour.

References

Aronowitz, S. (1978) 'Marx, Braverman and the logic of capital': Insurgent Sociologist, Vol. 8 Nos. 2 & 3, 126-46

Baggott, M. (1985) 'The vital support industries': 120 Scottish Economic Development Review, 4, 14-17

Bernstein, A., B. De Grasse, R. Grossman, C. Paine and L. Siegel, (1977) Silicon Valley: Paradise or

Paradox: Mountain View, Pacific Studies Center
Bluestone, B. and B. Harrison (1982) <u>The Deindustrialization of America</u>: New York, Basic Books
Borrus, M., J. Millstein and J. Zysman (1982) <u>International Competition in Advanced Industrial Sectors: Trade and Development in the Semiconductor Industry</u>, Joint Economic Committee, Congress of the United States, Washington, D.C., U.S. Government Printing Office
Braun, E. and S. Macdonald (1982) <u>Revolution in Miniature</u>: Cambridge, Cambridge University Press
Braverman, H. (1974) <u>Labor and Monopoly Capital</u>: New York, Monthly Review Press
Chen, E.K.Y. (1971) <u>The Electronics Industry of Hong Kong: An Analysis of its Growth</u>: M.Soc.Sc. Diss., University of Hong Kong
Cohen, R.B. (1981) 'The new international division of labor, multinational corporations and urban hierarchy' in M. Dear and A.J. Scott (eds) <u>Urbanization and Urban Planning in Capitalist Society</u>: London, Methuen, 287-315
Dunning, J.H. (1981) <u>International Production and the Multinational Enterprise</u>: London, Allen & Unwin
Edwards, R. (1979) <u>Contested Terrain</u>: London, Heinemann
Ernst, D. (ed.) (1980) <u>The New International Division of Labour, Technology and Underdevelopment</u>: Frankfurt, Campus Verlag
Ernst, D. (1983) <u>The Global Race in Microelectronics</u>: Frankfurt, Campus Verlag
Firn, J. and D. Roberts (1984) 'High-technology industries' in N. Hood and S. Young (eds) <u>Industry, Policy and the Scottish Economy</u>: Edinburgh, Edinburgh University Press, 288-325
Frank, A.G. (1980) <u>Crisis: In the World Economy</u>: London, Heinemann
Frank, A.G. (1981) <u>Crisis: In the Third World</u>: London, Heinemann
Friedman, A. (1977) <u>Industry and Labour</u>: London, Macmillan
Frobel, F., J. Heinrichs and O. Kreye (1980) <u>The New International Division of Labour</u>: Cambridge, Cambridge University Press
Gorz, A. (ed.) (1976) <u>The Division of Labour</u>: Brighton, Harvester Press
Grossman, R. (1979) 'Women's place in the integrated circuit': <u>South-East Asia Chronicle</u>, No. 66, 2-17
Hargrave, A. (1985) <u>Silicon Glen: Reality or Illusion?</u>: Edinburgh, Mainstream Publishing
Henderson, J. (1986) 'The new international division

of labour and American semiconductor production in Scotland': mimeo, Centre of Urban Studies and Urban Planning, University of Hong Kong

Henderson, J. and R. Cohen (1979) 'Capital and the work ethic': Monthly Review, Vol 31, No. 6. 11-26

Henderson, J. and R. Cohen (1982a) 'On the reproduction of the relations of production' in R. Forrest et al (eds) Urban Political Economy and Social Theory: Aldershot, Gower, 112-43

Henderson, J. and R. Cohen (1982b) 'The international restructuring of capital and labour: Britain and Hong Kong': Paper for the International Sociological Association's Xth World Congress, Mexico City, August

Hoogvelt, A.M.M. (1982) The Third World in Global Development: London, Macmillan

Hoover, E.M. (1937) Location Theory and the Shoe and Leather Industries: Cambridge, Mass., Harvard University Press

Hymer, S. (1979) 'The multinational corporation and the international division of labour' in S. Hymer The Multinational Corporation: A Radical Approach: Cambridge, Cambridge University Press, 140-64

Katz, N. and D.S. Kemnitzer (1982) 'Fast forward: the internationalization of Silicon Valley': mimeo, Department of Anthropology, San Francisco State University

Keller, J.F. (1981) The Production Worker in Electronics: Industrialization and Labor Development in California's Santa Clara Valley: Ph.D. Diss., University of Michigan

Lim, L.Y.C. (1978) 'Women workers in multinational corporations in Developing Countries: The case of the electronics industry in Malaysia and Singapore': Occasional Paper, Women's Studies Program, University of Michigan, Ann Arbor

Lim, L.Y.C. and E.F. Pang (1982) 'Vertical linkages and multinational enterprises in developing countries': World Development, Vol. 10 No. 7, 585-95

Locate in Scotland (1983) The Semiconductor Industry in Scotland: Glasgow, Scottish Development Agency

Marx, K. (1967) Capital Vol. 1: New York, International Publishers

Neff, R. (1982) 'Hong Kong takes a cut at technology': Electronics, March 24

Nichols, T. (ed.) (1980) Capital and Labour: Glasgow, Fontana

Noble, D. (1977) America by Design: New York, Oxford University Press

Norton, R.D. and J. Rees (1979) 'The product cycle and the spatial decentralization of American

manufacturing', Regional Studies, Vol. 13 No. 2, 141-
51
Paglaban, E. (1978) 'Philippines: Workers in the
export industry': Pacific Research , Vol. IX, Nos. 3
& 4
Rada, J. (1982) The Structure and Behavior of the
Semiconductor Industry: Geneva, UN Centre for
Transnational Corporations
Saxenian, A. (1981) 'Silicon chips and spatial
structure: the industrial basis of urbanization in
Santa Clara County, California': Working Paper, No.
345, Institute of Urban & Regional Development,
University of California, Berkeley
Saxenian, A. (1983a) 'The genesis of Silicon Valley':
Built Environment, Vol. 9 No. 1, 7-17
Saxenian, A. (1983b) 'The urban contradictions of
Silicon Valley: regional growth and the restructur-
ing of the semiconductor industry': International
Journal of Urban and Regional Research, Vol. 7, No.
2, 237-62
Sayer, A. (1982) 'Explanation in economic geography:
abstraction versus generalisation': Progress in
Human Geography, Vol. 6, No. 1, 68-88
Scott, A.J. (1980) The Urban Land Nexus and the
State: London, Pion
Scott, A.J. (1982) 'Production system dynamics and
metropolitan development': Annals of the Association
of American Geographers, Vol. 72, 185-200
Scott, A.J. (1983) 'Industrial organization and the
logic of intra-metropolitan location: I. Theoretical
considerations': Economic Geography, Vol. 59, No. 3,
233-50
Siegel, L. (1979) 'Microelectronics does little for
the Third World': Pacific Research, Vol. X, No. 2
Siegel, L. (1980) 'Delicate bonds: the global
semiconductor industry': Pacific Research, Vol. XI,
No. 1
Siegel, L. and R. Grossman (1978) 'Fairchild
assembles an Asian empire': Pacific Research, Vol.
IX, No. 2, 1-8
Siegel, L. and H. Borock (1982) Background Report on
Silicon Valley (Prepared for the U.S. Commission on
Civil Rights): Mountain View, Pacific Studies Center
Snow, R. (1980) 'The new international division of
labor and the US workforce: the case of the
electronics industry': mimeo, East-West Center,
Honolulu
Stigler, G. (1951) 'The division of labor is limited
by the extent of the market': Journal of Political
Economy, Vol. 59, 185-93
Thompson, P. (1983) The Nature of Work: London,

Macmillan

Tronti, M. (1972) 'Workers and capital': Telos No. 14, 25-62

Troutman, M. (1980) 'The semiconductor labor market in Silicon Valley: production wages and related issues': mimeo, Pacific Studies Center, Mountain View

United States Department of Commerce (1979) A Report on the U.S. Semiconductor Industry: Washington D.C.

Wallerstein, I. (1979) The Capitalist World Economy: Cambridge, Cambridge University Press

Wallerstein, I. (1983) Historical Capitalism: London, Verso

Wilson, R.W., P.K. Ashton and T.P. Egan (1980) Innovation, Competition, and Government Policy in the Semiconductor Industry: Lexington Mass., D.C. Heath

Chapter Four

**TECHNOLOGY WAVES AND THE FUTURE SOURCES OF EMPLOYMENT
AND WEALTH CREATION IN BRITAIN**

Paschal Preston

Introduction: The Scope and Context of this Contribution

What industries or economic activities are likely to
provide new sources of employment in the advanced
industrial economies such as Britain? Since the
ending of the long post-war economic boom and the
return of mass unemployment where can we expect to
find the means of reducing the dole queues? How can
we explain the recent and likely future changes in
employment structures in terms of contemporary
technological changes? To what extent will services
replace manufacturing industries as the sources of
new jobs? How do we best conceptualise the nature and
employment implications of current technological
changes? These are some of the key issues which have
concerned social scientists and policy-makers in
recent years and which this chapter will seek to
address.

Since the early 1970s, there has been a growing
body of literature within the social sciences and
elsewhere, focused on the issue of the likely sources
of employment, output and exports growth in the
advanced capitalist industrial economies such as
contemporary Britain. Here I will attempt to review
and select a number of diverse strands in this
literature and accompanying debates, in Britain and
relate them to recent empirical trends in the
structure of employment. The basic aim will be to
outline what, in the author's reading of the
literature and empirical evidence, appears to be the
most appropriate vantage point for conceptualising
the kinds of changes in the structure of employment
currently taking place in Britain, their
relationship to rapidly emerging and diffusing
technologies and identify the likely sectoral
composition of employment (1).

The contributions to the analysis of the recent and likely future changes in the structure of employment and output growth in the advanced economies such as Britain have been many in number and extremely wide-ranging in their particular sets of concerns and conceptual tools. Thus at the outset, we must stress and make explicit that we will here concern ourselves mainly with a select number of contributions which have primarily focused on technological change in their analysis of contemporary changes in employment and economic structures. We will attempt to evaluate and assess these in terms of their adequacy for understanding the recent observable and likely future structural changes in the British economy.

But in addition we will point to a number of concerns and contributions drawn from a body of literature which had little or no explicit focus on new waves of technology. Here we are referring to the debates over the phenomenon of 'de-industrialisation' which flourished in Britain in the late 1970s (see, for example, Blackaby, 1979). Although this literature often lacked the explicit concern with rapid technological changes characteristic of much of the recent contributions, it nevertheless sought to address relevant issues of structural changes such as the alleged shift to a 'post-industrial' economy, where services are said to become the major sources of output and employment in advanced economies such as Britain. This, as we will see, is a common element in some of the more technology-focused contributions. In addition, this literature is relevant in that it is more empirically-based and seeks to address comparative national trends and changes and the problems of Britain's long-run relative international decline as an industrial power which is germane to any attempt to locate the likely future sources of employment and wealth creation in the British space economy.

Thus in this chapter I will first of all outline one set of recent and influential contributions which seek to address the nature and implications of the current wave of new technologies for the structure of employment and economic activity in the advanced industrial economies in recent years. Having found these inadequate on a number of counts, the discussion will then proceed to outline a more adequate way of conceptualising the nature of the current new wave of technologies and their implications for the economic and employment structures. This will be based on recent

contributions to the literature dealing with the
phenomenon of Kondratieff long waves. In the light of
this, I will seek to examine some of the recent
sectoral changes in the structure of employment in
Britain and their implications for the future
patterns of economic activity.

At the outset, it must be pointed out that
although this contribution adopts a strong focus on
technology as a force underlying change in economic
activities and employment structures, this does not
imply that technology is an all determining or
autonomous factor. On the contrary, technology must
itself be conceptualised as largely the product of
economic and social relations. Thus the past and
likely future changes in the structure of employment
are more a function of the social, economic and
indeed political relations in which particular forms
of technology have developed and are diffusing, than
any inherent function or 'logic' of the new
technologies themselves.

A **Third Technological Revolution?**

This section will outline and evaluate one particular
and widely-held perspective on the nature of
contemporary technological changes and their
implications for the future structure of economic
activities and employment in the advanced industrial
economies. Basically it suggests that the nature of
contemporary technological changes is such that it
represents nothing less than a third technological
revolution in the history of human society. It
suggests that the significance of the technological
changes currently taking place must be viewed in
historical terms as of the same order as the
agricultural revolution of 10,000 years ago and the
industrial revolution of 200 years ago, each of which
brought about total transformations in the existing
forms of economic, employment, social, political and
family life and activities. As one proponent puts it
a new critical transition is upon us, as contemporary
trends in technology, economics, politics, family
life, energy use and other spheres of life usher in a
third civilisational rupture - a transition to Third
Wave social forms (Toffler, 1983, page 11).

This 'third wave' or 'third technological
revolution' perspective on contemporary technological changes can be found in many discussions and
analyses of their implications for the structures of
employment and economic activities in the advanced
industrial economies, both amongst social scientists

and policy makers. Basically it suggests that the sources of employment and output growth in the immediate future will be radically different to those that have prevailed over the past couple of centuries. In particular it suggests that the vast majority of jobs will be concentrated in the higher level services, or information producing and handling or 'service' occupations unrelated to manufacturing industries. In this author's estimation, this perspective is not a helpful one when it comes to identifying the likely sources of employment growth for a number of reasons that will be made more apparent by discussing the contributions of Toffler, (1980, 1983) and Stonier (1983) two authors who adopt such a perspective.

(a) **Stonier's Contribution.** In Stonier's view, the contemporary processes of technological innovation and the economic and social changes associated with the relative growth of services and information occupations and the new information technologies are of the same order and historical importance as the industrial revolution at the beginning of the nineteenth century when Britain was transformed from an agricultural economy and society to a 'post-agricultural' one. He says:

> In Queen Victoria's heyday, under the pressure of advancing technology, Great Britain had shifted to an industrial economy ... Today we have witnessed a similar shift. From a major exporter of manufactured goods, the United Kingdom has become an importer ... Only a shrinking minority of the labour force toils in factories; and the service sector has overtaken manufacture in terms of the country's gross national product. We now live in a post-industrial economy. (Stonier, 1983, page 7.)

So what are the key economic activities and sources of employment and wealth creation in this radically new socio-economic system? According to Stonier, in a post-industrial society, a country's store of information is its principal asset, its greatest potential source of wealth (ibid, page 12). He argues that in the advanced industrial world, information has 'upstaged' land, labour and capital as the most important input into modern productive systems. He says that information reduces the requirements for land, labour and capital, raw materials and energy and that it spawns entirely new industries. He argues that information is increasingly sold in its own right and that it is the raw material for the fastest-growing sector of the economy, the knowledge industries.

In Stonier's view, 'the labour market is now dominated by information operatives who make their living by virtue of the fact that they possess the information needed to get things done' (ibid, page 8). He criticises classical economists and especially contemporary economists for their failure to recognise the growing importance of knowledge and information as an economic resource. He suggests that conventional economic theories cannot grasp the apparently new phenomenon that the conversion of non-resources into useful products has become the basic principle of new wealth creation because of the advanced state of technology and knowledge in post-industrial society.

Stonier stresses that wealth creation is now increasingly based upon advanced technological expertise, and dependent upon a base of information and knowledge, especially advanced science, rather than any other resources or factors. He views scientific activity and advances as playing an increasingly autonomous and influential role in the development of the economic system, 'unlike earlier times when practical experience laid the foundations for new science, the flow of information is now in the opposite direction' (ibid, page 12).

So where will we find the new sources of employment, wealth and export earnings in Stonier's post-industrial system? In his book there appear to be a number of conflicting answers to this question, but the general thrust of his argument is as follows. He claims that within 25 years, it will take no more than 10% of the labour force to provide us with all our material needs (ibid., page 122). Labour has begun to shift to the new, technologically advanced, information-intensive productive systems which require not only a quantitative reduction in the workforce but also qualitative changes such as much greater inputs of mental effort and reductions in manual effort. In this view, as information inputs continue to expand, shop floor workers will require advanced engineering degrees, just as American farmers require a university degree to take advantage of his technological back-up (ibid, page 122). So even the 10% of the workforce needed in agriculture and manufacturing etc. to provide all society's clothing, food, housing, transportation, and appliances will become more and more involved with handling complex information.

Stonier's 'third-wave' perspective suggests then, that in Britain's emerging post-industrial economy most workers will be people who spend most of their time manipulating information and getting things started, or alternatively people working with people as in education, health or other caring activities. Basically his scenario and related economic strategy proposals suggests that Britain's comparative advantage and future export earnings do not lie in manufacturing activities, but in a combination of two other sets of activities. One is a revived, post-industrial, primary sector whereby agriculture and mining, originally pre-industrial activities, rise once again to importance; the second consists of a knowledge-related services sector producing business-related and professional services, a 'post-tertiary sector' making invisible earnings in international banking, insurance, shipping, air traffic and tourism.

Stonier claims that one of the 'ironies' of the post-industrial economy is the revival of the primary sector as a major source of employment and export earnings, citing the recent cases of oil exports from Britain and agricultural exports from the USA, which have helped replace jobs and export earnings lost in manufacturing. He discusses at length four major sets of potential technological developments, which if properly developed, could make Britain both energy independent after the oil runs out and a net exporter of food and chemicals. These concern: the conversion of coal to oil, gas and chemicals; developing wave powered electricity generators; coastal fish-farming; and single cell protein for livestock feed. He also discusses at some length the potential for developing other new primary sector technologies such as ocean farming, three dimensional farming and deep-sea mining. These could be the sources of new wealth and economic activity for Britain, even if their potential applications are located in other parts of the world; for many of these developments, he argues that the limiting factor in 1980 was not so much the technology or the economic potential, but international law. Who owned these sites? (ibid, page 63).

However, his vision of a growing primary sector also includes more mundane and familiar, not to mention mature, products such as cheese and biscuits. On a number of occasions he waxes lyrical about the superb quality of British cheeses and the light, dry biscuits to go with the cheeses, however the trouble is that the world does not know about them (ibid, page 169); but with proper marketing, cheese products could be a major export item, especially when single cell protein for feeding dairy herds becomes a viable technology (ibid).

His optimism about the primary sector is not shared by manufacturing industries, however. Overall he has a rather negative, if at times inconsistent, view of its future role as provider of employment, output and export growth. Towards the beginning of the book he says that like British agriculture before it, manufacturing will continue to be important in terms of output, but will decline rapidly as a source of employment in the post industrial economy (e.g. pages 23-24). Towards the end of the book, however, he suggests that it may not even matter if it declines in terms of output (e.g. pages 145-148). Indeed at once point he declares that there is no need to stay in the manufacturing game at all (e.g. page 147). Indeed he appears to dismiss and/or neglect the potential wealth, employment and export potential of the new information technology producing industries.

Thus in this perspective, apart from the potential revival of primary-sector activities, the major sources of new employment and wealth are to be found in the information/knowledge producing industries, in Britain's capacity to produce or generate knowledge and provide certain kinds of information. He claims that it is in the knowledge industry that Britain has its greatest long-term potential; and that if during the nineteenth and most of the twentieth centuries, Britain's global economic role was that of industrial machine shop, it must now become its post-industrial, technical-managerial consultancy and information provider (ibid, page 148).

What kind of specific activities will this entail? Stonier argues that over the next few decades, education will grow to become the largest single industry of the post-industrial society and its number one employer. He believes that in a rapidly-changing information society, people will need to move in and out of education throughout their lives and we will need a whole new cradle-to-grave system which includes community education for the mature. He stresses that universities and other higher education institutions, do not simply provide employment and improve human capital, but they also play a crucial role in producing new ideas and new

industries and that a massive expansion of the
education system must be the cornerstone of any
government policy to facilitate the transition to a
post-industrial society in order to provide the
skilled workforce required by an information
economy. He suggests that a doubled education budget
would give employment to more than a million people
and would be self-financing by cutting unemployment
benefit costs and would generate foreign exchange
earnings by providing for overseas students.

On a related point Stonier stresses the
importance of a growing role for research and
development activities in wealth creation. He argues
that R&D activities have become increasingly
important in quantitative employment terms, but that
they are also highly institutionalised and require
direct government finance in order to ensure the
adequacy and continuity of research effort.

In addition to these, Stonier stresses the
importance of market-orientated information and
knowledge industries. He suggests that these should
build upon the City of London's long history as an
international centre for banking, finance, insurance
and shipping. He stresses that its telecommunications linkages and air transport networks to other
parts of the world are amongst the best in the world,
and like education services, one of its greatest
assets is that its native language is English. The
scope for developing London into the office
activities capital of the world could be enhanced by
the development of all sorts of cultural and leisure
industries, ranging from theatres and museums to
night clubs and the amorous activities of Soho or
Shepherds Market (ibid, page 163).

If global business information and office
activities will provide the jobs for London and the
South East, the equivalent wealth and job generation
activities for the remainder of the country will lie
in tourism according to Stonier. He stresses that if
this industry is erratic, it has now become a bigger
industry in Britain than automobiles, and with
Britain's historical heritage, not least in its
relics of the industrial society era, there is scope
for further expansion. With careful planning and
government finance and support, he suggests that the
North of England is ripe for educational tourism,
centred on the history of industrialism (ibid, page
166).

Whilst Stonier's contribution makes some
important points concerning the growing importance
of information as an economic input, and his

discussion of potentially emerging technologies of the more distant future is highly accessible and informative to the non-technologist, overall it provides an inadequate guide to the nature of the currently emerging technologies and their implications for the structure of employment in the medium term future as we will demonstrate below. But first we must consider the contributions of Toffler.

(b) **Toffler's Contribution.** Alvin Toffler is one of the most widely read and influential commentators on technological and socio-economic change in the contemporary world. His books, Future Shock and The Third Wave have been international best sellers and have been translated into many languages. His newspaper articles and television programmes on this subject have reached a wide audience in many countries.

Toffler's main theme in The Third Wave (1980) and other more recent works such as Previews and Premises (1983) is that the world industrial system, both capitalist and socialist, is undergoing a process of fundamental restructuring or 'general crisis'. He argues, much like Stonier, that contemporary changes are of the same order and significance as the agricultural revolution of 10,000 years ago and the industrial revolutions beginning 250-300 years ago, each of which brought massive transformations to existing forms of social, economic, political, family, etc. life.

Referring to the decline of the mass manufacturing industries, such as steel, autos, textiles, and the simultaneous rise of electronics, information, aerospace, etc., he claims that what is happening today is not a recession as such, but a restructuring of the entire technoeconomic base of the society. He suggests that 'it's like an earthquake that throws up a new terrain ... in short-hand terms, we're shifting from a Second Wave to a Third Wave economy (1983, pp.11-12). He says that the central proposition of his work, as it effects economics, is that today's crisis is not a crisis of the re-distribution, or of over- or under-production, or of low productivity but 'a crisis of restructure', - the breakdown of the old Second Wave industrial era and the emergence of a new Third Wave economy that operates on entirely different principles.

What are key features of the changes associated with new technologies which Toffler observes taking place today? Firstly, he says that the changes are essentially global in scope and affect most, if not all, countries. Secondly, he stresses that the crisis and restructuring processes cannot be viewed as either 'capitalist' or 'communist' but industrial; indeed he stresses that the revolutionary nature of the new technologies is such that they imply a total transformation of all existing economic social and political relations and conflicts; they mean that the basic structure of social differentiation and political conflict has shifted away from that of social classes such as capitalists and workers to that of the undifferentiated vested interests of workers, employers and politicians committed to the 'Second Wave' industrial system versus those committed to the 'Third Wave' system. He says that in this historic 'super-struggle' and clash of civilisations, the 'basic groupings don't line up with class, race, sex, or religion, or with left or right, or with existing political parties' (1983, p.89).

A third key feature of the new technologies and associated emerging economic system is that it involves a shift from a mass production and mass consumption economy to a 'de-massified' one. In his view of the rapidly developing 'Third Wave' sector, mass production is replaced by its exact opposite, short-runs, or even customised one by one production, based on automated systems which he claims make diversity as cheap as uniformity. A fourth feature of Toffler's Third Wave economy is that it contains a significant and growing sector based on what he calls 'prosuming'. Here he is referring to the processes whereby people produce more goods and services not for sale, or even for barter, but for their own use. In his view, this growth of production for consumption by the producer fundamentally alters the relationship of the consumer to the production process. Here we may note that Toffler's concept of 'prosuming' is similar in many respects to the notion of the 'self-service' economy as a key feature of post-war social innovation, which has been identified in a less sensational and more empirically-informed manner by Gershuny (Gershuny, 1983; Gershuny and Miles, 1983).

A fifth feature of Toffler's argument is that today, national economies as such may become outmoded. This he claims, is not simply because a higher and higher percentage of world production takes place for multinational as opposed to national markets, but because of an 'opposite trend that has

been gaining momentum at the same time - the shift
from national to smaller-than-national production'
(page 19). He believes that regional economies are
growing more and more divergent rather than
convergent and this partly explains what he sees as a
rising regionalism in culture and politics. Toffler
suggests that these trends will mean a greater degree
of regional autonomy and independence as 'managing
their economies centrally, from the national level
... will become more and more difficult' (1983,
p.20). His discussion of this apparent phenomenon is
closely bound-up with his claims about increasing
social and individual diversity arising from the new
technologies. This is related to a sixth feature of
Toffler's Third Wave society whereby the new
technologies and associated innovations will give
rise to new organisational forms of the most diverse
kinds. He says that we will find 'electronic co-ops',
religious and familial production teams, non-profit
work networks and many more novel arrangements than
we can now imagine, including a major shift to home
working. He says that 'mind-workers' will be far
better equipped to manage themselves than typical
workers of the past and indeed he suggests that they
will really own and control the means of production
since 'in a third wave society ... the essential
property becomes information' (1983, p. 103).

So what are the specific forms of economic
activity and sources of employment that Toffler
envisages for our immediate future? He stresses that
these cannot be viewed as an extension of, or simple
development from existing 'Second Wave' industries.
He argues that they are qualitatively different in
many significant respects: '... they differ in a
thousand ways. The kind of product. The kind of
people in them. Their organisational structures.
Their style and culture. And at the most profound
level - the level of knowledge - they represent a
fundamental break with the past' (ibid, p.17). He
emphasises that they imply a wholly different kind of
work, with each worker performing many different and
complex operations, participating in key decision-
making, working in clean and quiet surroundings and
operating on flexi-time instead of rigid hours. All
workers will have to acquire a wide range of skills
and specialised knowledge and cultural values on a
continuous basis. This means that the future high-
tech society will have to provide continuous and
intensive training and education programmes, and
like Stonier, Toffler suggests that education and
training will themselves constitute very major

activities and sources of employment.

He rules out any future major employment role for manufacturing industries, at least those involved in mass production. But he suggests that some may survive, if they drastically restructure in both technological and organisational terms or for strategic and military reasons. He is sharply critical of those, such as unions or governments which seek to maintain the 'sunset' or 'rustbowl' manufacturing industries. Much like Stonier, he seems to point to knowledge and information producing or handling activities as the major source of employment, but he seems generally unwilling to specify the new forms of economic activity. He stresses that we can only begin to appreciate the full impact of the new technologies when we relate them to his visions of (i) new 'de-massified' production and distribution processes (ii) the altered roles of producers, consumers and third parties in the economy and (iii) when they are linked together in convergent systems. But at one point he suggests that the new 'Third Wave' industries will range from 'electronics, lasers, optics, communications, and information to genetics, alternative energy, ocean science, and space manufacture, ecological engineering and eco-system agriculture, all reflecting the qualitative leap in human knowledge which is now being translated into the everyday economy' (ibid., p.18).

At one point he also warns that we may have to face the fact that not everyone can have a 'job', in the sense of a formal, paid, productive job in the exchange economy. He suggests that eventually we may need to explore 'prosuming' ways in which people can produce for themselves, rather than for sale or barter. He also suggests minimum income guarantees as part of 'the moral basis for a totally new, humane system of rewards that match the fresh options opened for us by a Third Wave economy' (ibid., page 59).

A Critique of 'Third Wave' Analyses
This section will seek to identify a number of theoretical and conceptual weaknesses inherent in the 'third wave' or 'third technological revolution' conceptualisations of the nature and implications of contemporary technological changes that were outlined above. It will also point the way to a more coherent and realistic approach outlined in the next section (See Preston, 1984, for an extended evaluation). Despite some of the clear advantages and

insights offered by these 'third wave' approaches to the analysis of contemporary structural, technolog- ical and socio-economic change compared to many conventional economic analyses, they do however, appear to be erroneous in a number of respects, including the following.

First, we should note that there is a certain temporal ambiguity or inconsistency in their analysis; it is not always clear whether they are referring to changes that have already occurred or are actually occurring now on the one hand, or changes that are about to occur in the future on the other hand. For example, Toffler at times seems to suggest that some existing 'high-tech' economies have already become Third Wave societies, whilst at other times it is obvious that he is referring to futuristic socio-economic-technical systems very far advanced from anything that exists today. The question and issue of defining the temporal context is not just a case of academic nit-picking. If they are referring to long distance futures, then it may well be the case that both Toffler and Stonier are correct in arguing that there is no point in undertaking restructuring initiatives to improve output and exports (and perhaps even employment) in the existing manufacturing or 'the Second Wave' sector. However, if we are concerned with the present and next say, 20-30 years, then this approach amounts to utopian nonsense. This is a failing shared with much of the 'post-industrial' society literature of a decade ago surrounding the publication of Daniel Bell's (1973) contribution and stressed by Newberg's (1975) review of these debates.

On balance, their analysis and discussion tends to be predominantly focused on the long-term future possibilities of our socio-economic and technical capacity. Whilst I believe that such discussions are stimulating and they give an extremely optimistic and positive view of our possible futures, they tend to have very little to say about the 'nuts and bolts' (or is it 'chips and bits') of how we get there by starting to tackle the pressing unemployment problems of today and tomorrow morning. But they are more confusing than helpful in that they conflate and mix-up discussions of what are currently existing and rapidly diffusing technologies on the one hand, with those technologies that are still very much in their embryonic stages on the other hand.

Secondly, they lack an adequate historical analysis of the nature of past phases of deep technological and associated socio-economic changes with the result that they tend to exaggerate the nature, impact and implications of present-day technological and associated changes. In part, this results from a failure to adequately analyse the long-term historical trends towards the development of an information-based/knowledge-based economy since the industrial revolution. Admittedly, as several recent studies have emphasised (e.g. Gershuny and Miles 1983; OECD 1981; Hall and Preston 1985; Porat 1977), these changes have often been masked by traditional approaches to and classifications of, industrial and employment data. However, the shifts are not as recent, not as dramatic in their effects, as Toffler and Stonier seem to assume. But there are other reasons to doubt the historical-basis and accuracy of Toffler's 'Third Wave' (or long-long wave) approach. The major one is that for him the uniqueness and 'Third Wave' technological changes is that at the same time as established, basic industries are experiencing slump, layoffs, over-production etc., a batch of new industries are simultaneously rising up and expanding. He implies that this has not happened before in industrial/capitalist societies and this is a major reason for regarding present day innovation processes as very different to any other that have occurred since the beginning of the industrial revolution (e.g. 1983, pp.10-11). This is nonsense for anyone with even a basic grasp of industrial history. For this reason alone one could suggest that the contemporary processes of innovation are best viewed in terms of Kondratieff long-wave framework rather than that of a unique or fundamentally new technological revolution as suggested by Toffler's Third Wave conception.

A third point worth noting about these analyses is their tendency towards technological determinism. This is probably more true of Toffler than of Stonier - who at least attempts to relate, more consistently, the social and policy context of innovation, to his discussion of the intrinsic qualities of the new technologies and their implications. However, both writers, it appears to me, place too much emphasis on the latter and tend to derive social and political impacts in an unwarranted fashion. See for example their discussions of the intrinsic, abstract qualities of information as a resource and their political and social effects, in Stonier (1983, pp. 18-19) and Toffler (1983, pp. 22, 16-22, 35-36, 100-104). A fourth point which is closely related to the other three just mentioned, is that alternative, conflicting political values and perspectives and policy options are effectively ignored or dismissed. We are simply told that value conflicts between 'left' or socialist parties versus 'right' or conservative ones concerning strategies for economic expansion, unemployment, welfare services, social inequalities or renewal of existing infrastructures, not to mention strategies for industrial restructuring, are now redundant in the fact of the new dawn accompanying the 'sunrise' industries apparently looming on the horizon.

A fifth point is that it is very difficult to see the evidence for Toffler's claim that alongside the growing multinationalisation of production, there has been, gathering momentum at the same time, the shift from national to smaller-than-national production (1983, p.19), and a rise of regional diversity and autonomy in terms of economics, culture and politics. It is very difficult to understand the nature and basis of this claim. Like Stonier (1983) and Radice (1984), we would argue that if nations as economic and cultured entities are disintegrating, it is overwhelmingly in the direction of internationalisation (or supra-nationalisation), rather than towards sub-national units. In the current slump, the internationalisation of economic relations has increased significantly and there has been a growing penetration of multinational corporations into the new technology and related service industries (Andreff, 1984). A sixth criticism is that it is also highly questionable that we have reached or are about to reach a stage of production where goods and other system outputs can be accurately described as 'de-massified' or highly diverse. Even in the archetypal new, 'Third Wave' product markets, such as that of micro-computers, the scope for such diversification is clearly very, very limited indeed; standardisation of operation systems and applications programmes and 'IBM-compatibility' is the name of the game.

Towards a More Adequate Long Waves Perspective on Current Technological and Economic Changes

In this section I will attempt to outline what I consider to be a more adequate and coherent way of conceptualising the nature of contemporary technological changes and of grasping their implications for the sources of wealth and employment creation in the near and medium term future. This is based on a particular reading of the growing literature on Kondratieff, long-wave cycles of economic growth and their relation to technological changes (e.g. Freeman, Clarke and Soete, 1982; Freeman 1984; Mandel 1975; Mandel 1980; Perez 1983, 1985; Gershuny and Miles, 1983; Kaplinsky, 1985; Van Duijn, 1983; Mensch, 1979).

The phenomenon of 'long wave' economic cycles of approximately 50 years duration since the industrial revolution was first identified in the first two decades of this century by van Gelderen, de Wolff and Kondratieff (see van Duijn, 1983, for an excellent history). But it was Schumpeter, in his major study of business cycles published in 1939, who developed a comprehensive account linking these long run fluctuations to technological changes and innovation. After the Second World War and during the great bulk of the 1950s and 1960s, there was very little interest in this perspective on technological and socio-economic change. That is until the past few years when the end of the boom, the return of mass unemployment and the slowdown in economic growth has been accompanied by a mini-boom of interest and publications based on this perspective.

Both the early and the more recent literature on long waves has embraced a number of different and conflicting positions concerning the precise determinants and implications of the fluctuations in growth and the role of technological change. One major difference is that between those who view phases of innovation and rapid economic growth as endogenous to the economic system, the result of 'demand-pull' forces and those who view it as exogenously 'driven' by technological changes. Proponents of the demand-pull thesis essentially view technical change, innovation and associated economic growth as passively responding to the 'carrot' of increases and changes in the patterns of demand. Technical change is afforded little or no independent influence on economic fluctuations. On the other side, there are those who focus on an apparent bunching of technological inventions and innovations during the depression phases and stress their influence in pushing the economy towards a new phase of economic growth (e.g. Mensch, 1979). In line with a number of other recent contributions, however, I believe that it is possible to transcend these polarised positions and adopt an alternative approach. This both recognises the relative autonomy and externality of initial technological changes to purely economic demand influences, but also recognises that demand patterns, along with other changes in economic conditions such as

profitability, as well as the role of the state, can
and do play an important role in the important
processes of diffusion and swarming of new technology
systems (Freeman et al, 1982; van Duijn 1983; Dosi,
1984; Perez, 1983; Mandel, 1980).

This approach stresses the importance of both
the emergence and rapid diffusion of combinations or
related sets of individual product and process
innovations (which constitute a new technology
system or paradigm) as well as innovations in the
organisational, managerial and institutional environment in leading to a phase of rapid economic
growth. In other words, technological change plays an
important role in the timing and location of phases
of rapid economic growth, but it is not the only
determinant. The rapid diffusion and swarming of
innovations requires other appropriate conditions as
well. This suggests that the long-run growth of
scientific and technological inputs into the
industrial system means that in the present era,
especially, the performance of any economy is
strongly influenced by its levels of R&D, scientific
and related educational provision. This approach
overcomes the problem of technological determinism
inherent in a number of other long wave and indeed
'Third Wave' perspectives. It stresses that 'since
technology is the "how" and "what" of production, it
is in fact very much a social and economic matter'
(Perez, 1985, p.442). It suggests that whilst the
process of technological advance in terms of
knowledge and inventions is a relatively autonomous
process, innovation, i.e. the application and
diffusion of specific techniques in the productive
sphere, is very much determined by social conditions
and economic profit decisions.

Another related difference amongst the
contributors to the long waves literature concerns
the emphasis they give to the apparent recurring
cyclical features of every long wave on the one hand
or to the specific characteristics of each particular
long wave on the other (Rosenberg and Frischtak,
1984). In brief, the former group tend to focus on
the various statistical measures and indicators of
the more quantifiable aspects of the innovation
process and adhere to the notion of a fixed,
recurring cycle of specific duration (e.g. Mensch,
1979). They also neglect social and institutional
aspects of the socio-economic system such as the role
of the state. The latter group tend to downplay or
reject the necessity of rigidly fixed statistical
cycles preferring instead to stress the complex

determination of socio-economic life and the variable combinations and influences which will cause the timings and extent of the fluctuations to vary from country to country or between waves. (Freeman et al, 1982; Perez, 1983). The latter also tend to focus on the many dominant or leading organisational, managerial, institutional and social innovations that are deemed most appropriate for each long wave new technology system; they stress that each long wave expansionary phase is focused around a new technology system or style and a related set of changes in key elements of the total socio-economic system. They suggest that 'a dynamic harmony is established among the different spheres of the total system' (Perez 1983; page 358).

In adopting this latter focus it is possible to identify how each long wave new technology system is associated not only with leading new technology based industries, but also with particular dominant forms or 'ideal types' of organisation of the production process, market and occupational structures, styles of management-labour relations, and forms of state intervention. It also allows us to examine not only changes in the location of innovation, but also changes in the dominant spatial scales/levels of economic relations and in the particular forms of the spatial division of labour (Preston, 1985; Perez, 1983).

This particular long waves perspective permits a more plausible and coherent approach to the analysis of contemporary technological changes and their associated implications for the form and nature of economic activities which might provide the means of reducing the current high levels of unemployment in the immediate future. In the first place it is more concretely rooted in a historical analysis of technological change and the emergence of new industries in the development of capitalist societies than that offered by 'Third Wave' perspectives. It is firmly rooted in the analysis of previous fluctuations of economic growth and employment levels and it is informed by the historical facts that the declining importance of mature industries in the past was compensated, at least in part, by the rise of more technologically advanced industries and a range of new social and institutional innovations and structural changes. Thus the current slump and associated mass unemployment and the emergence and availability of a range of new technologies with a potentially wide application in terms of product, process,

institutional and organisational innovations is far
from an entirely novel phenomenon in the history of
capitalism. Although the specific forms of the new
technologies, their potential applications and
impacts are different to those of the past, and
although history can never repeat itself, we ignore
it at great cost in this type of analysis. Although
we might agree with Toffler and others who suggest
that the speed and intensity of contemporary
technological changes are different to those of the
past, this in itself does not justify a rejection of
the long wave approach in favour of some 'Third Wave'
perspective. Within the former approach, it is
possible to recognise, for example, that there was a
significant qualitative and quantitative deepening
of the scientific and technological basis of
industrial activities during the third long wave,
beginning in the 1890s, compared to that of the first
two long waves (Freeman et al, 1982; Hobsbawm, 1969).
This prompted contemporary and subsequent observers
to perceive it as a 'second' industrial revolution
much in line with Toffler's and Stonier's perspective
today. But as Hobsbawm points out, as time goes on
the '"second revolution" becomes assimilated to the
past' and in due course 'another "second" industrial
revolution is discovered - maybe in the 1920s, and
then again in the age of experiments in ambitious
automation after the Second World War' (1969, p.172).
   Thus the historically informed long waves
perspective outlined above offers a more coherent,
realistic and less sensationalist perspective on the
nature of contemporary technological changes and
their implications for changes in economic and
employment structures than the third wave approaches
discussed earlier. In addition, because the long wave
perspective outlined above is more rooted in the
historical developments and empirically-informed, it
tends to be more cognisant of the pace and pattern of
technological changes and in particular it is mindful
of the long-run growth of information/knowledge and
associated service activities within the advanced
capitalist industrial economies (e.g. Gershuny and
Miles, 1983). Furthermore, because of its stress on
the importance of diffusion and 'swarming' of
existing or emerging technologies in facilitating
phases of rapid economic growth, it is more
concretely focused on likely developments in the
immediate future. It thus helps to avoid the
confusion and conflation of these with more embryonic
and distant developments as usually occurs in 'Third
Wave' analysis.

In addition the long waves perspective outlined
above makes it possible to avoid some of the other
points of criticism of the 'Third Wave' analyses
mentioned earlier. For example, it avoids the problem
of technological determinism by allowing for the
importance of institutional, organisational innovations and indeed the role of the state in shaping
the form, extent, pace and timing of their
development, diffusion and impacts (Freeman et al,
1982; Perez, 1983; 1985).

For these reasons then, I believe that the long
waves perspective outlined above provides the most
helpful coherent approach to understanding the
nature of contemporary technological changes and
conceptualising their implications for the likely
forms of employment and economic activity in
contemporary Britain.

The Future Sources of Employment and Wealth Creation in Britain: A Long Waves and Empirical Analysis

So what are the likely forms of economic activities
and types of employment in advanced economies such as
Britain within the long wave perspective outlined
above? How do these differ from the types of
activities suggested by writers such as Toffler and
Stonier? Here it must be stressed that there is no
single simple answer to this question in the
particular set of this literature that the present
author has examined and which informs the long wave
perspective just outlined. There are many
differences between these authors concerning, for
example, the likely geographical changes in the
international location of new and existing
industries, the extent to which the newly emerging
and diffusing technologies will imply changes in the
organisational (enterprise) structures of industry,
the implications for skill structures and so on. To
some extent this uncertainty is to be expected and
even welcomed for it is based on a firm recognition
of the dangers and limits of reading-off too much
from purely technological changes and even currently
widely diffusing ones. It reflects the necessity of
recognising that the precise impacts and
implications will be shaped by a myriad of system-
wide factors including social and political
responses by the key groups of actors both in the
economic and political spheres. Despite these
differences, however, this body of literature
provides a more useful starting point for such
analysis and by taking account of recent empirical

evidence, it enables us to get a better picture of likely patterns of employment in Britain in the immediate future compared to the Third Wave perspectives as I will now attempt to demonstrate.

Firstly, because of its common focus on the importance of currently existing and potentially widely diffusing technologies, within this body of literature there is a general concensus concerning a more precise identification of the leading-edge, new technology-producing industries of the present and immediate future. These are generally agreed to be the electronics-based new information technology (NIT) industries. Rather than embracing the more futuristic developments over which writers like Toffler and Stonier wax lyrical, the long wave theorists generally focus on the convergence of the new technological advances in microelectronics, computing and telecommunications as providing the key technological infrastructures of the present and immediately future decades.

These new IT industries, supplying a range of new products for final consumers as well as inputs for a range of other activities may thus be viewed as key growth industries for the advanced industrial economies. As the leading, new technology supplying industries of the current era, they would be expected to expand in employment, output and exports terms even during the current downswing, just as previous leading-edge industries did in the past, for example, the vehicle and aircraft and electrical producing industries during the 1930s slump.

How does this expectation relate to concrete empirical evidence on recent economic and employment changes in the British economy? In adopting a fairly broad definition of these new IT industries in terms of the official Standard Industrial Classification, we find that between 1971 and 1981, whilst these industries experienced a growth in output they actually experienced a drop in employment of about 12% over the same period (see Preston et al, 1985 for details). Does this mean that the long-wave perspective and the expectation that these new IT industries would provide additional sources of employment in the developed economies should be discarded? I believe not, partly because the empirical evidence for a range of other countries confirms the hypothesis. For example, the Reading University international data base on these industries shows that employment increased over the same period in all but three out of a group of 26 OECD-member and newly-industrialising countries. Ir

Japan employment increased by 25% and in the USA by 43% in these industries over the same period. Thus the expectation that they should also constitute an additional source of employment in Britain is a valid one. The fact that they don't highlights, rather than undermines, the value of the perspective we have adopted and it serves to underline the relatively poor innovative performance of the British industrial system during the fourth Kondratieff long wave and especially since the downswing from 1973 (Pollard 1984; NEDO, 1984). The failure of the IT industries in Britain to generate additional employment since the early 1970s reflects wider economic, social and political processes and serves as a timely reminder of the need to relate technology-centred analyses to both concrete empirical analysis and wider socio-economic contexts.

Secondly, in addressing the question of the skill and occupational composition of employment associated with the new technologies, and empirically-informed long waves perspective gives us a very different picture to the one presented by Third-Wave writers such as Toffler and Stonier. It will be remembered that the latter have suggested that the majority if not all of the jobs in the new 'third wave' society will be concerned with handling and producing information/knowledge, and that even manufacturing employees will require degree-level skills and qualifications. In this regard, too such analyses present a very misleading picture of the actual and potential future trends in the skill structure of employment. This can be made more clear by referring to the occupational structures of the new IT producing industries themselves. As technology-intensive, leading-edge industries, they might be expected to have relatively high proportions of research, scientific and technical staff compared to other 'mature' industries. These might be taken as pre-figurative of the kinds of wider changes in the occupational structures which might be associated with contemporary technological changes. As the industries producing the new information technologies, they are themselves the most intensive users of the new technologies in their own production processes and as such their occupational structures might be regarded as typical of manufacturing in the future decades.

When we examine the recent occupational data for a relatively broad group of the NIT industries in Britain, we find that they do indeed have relatively

Table 4.1: Employment by Occupation in Selected New IT Industries, 1978 and 1984 (GB)

Occupational Category (1)	All Employees 1978		All Employees 1984	
	No of employees (2)	Proportion of total employment (3)	No of employees (4)	Proportion of total employment (5)
Managerial staff	20,513	5.2	21,505	6.4
Scientists and technologists	20,914	5.3	34,512	10.3
Technicians including draughtsmen	44,464	11.2	44,802	13.4
Administrative & professional staff	28,761	7.3	31,633	9.5
Clerks, office machine operators, secretaries and typists	51,282	13.0	39,860	11.9
Supervisors including foremen	21,353	5.4	15,854	4.7
Craftsmen	29,019	7.3	22,632	6.8
Operators Other employees	179,325	45.3	123,258	37.0
Total employment	395,631	100.0	334,056	100.0
Male Total Employment	240,753	60.9	219,162	65.6
Female Total Employment	154,878	39.1	114,894	34.4

Note: This data relates to SIC categories 3301, 3302, 3441, 3442, 3443, 3444, 3453, 3454.

Source: EITB Statutory Returns.

high proportions of scientific, technologist and technical occupations. These occupations have increased their share of total employment from about 17% to 23% between 1978 and 1984 (See Table 4.1). We must, however, bear in mind that about half of the workers in these exceptional industries are still in the more routine, low-skill occupational groups of 'operators' and 'clerks etc.' Thus the empirical data for even these technology-intensive 'sunrise' industries in Britain, does not suggest that we are about to embark on the people-less, automated manufacturing production systems run only by a small number of engineers – at least not in the immediate future.

Here again, the sensationalised and futuristic picture painted by the third-wave theorists must be replaced by a more sober and historically informed view of the extent and implications of the current wave of automation technologies. This point is even more important when we remember that many of the NIT industries currently have highly internationalised production systems with many of the labour-intensive functions taking place in less developed countries. If the skill structures, wage levels and working conditions of the NIT sunrise industries are analysed fully in their international context, then the shifts in the nature of work and occupational structures implied by the new technologies is indeed even less rosy and optimistic than that suggested by writers such as Toffler and Stonier. (Bowlby and Preston, 1985.)

The new automation systems which are developing on the basis of the new information technologies of today, will not mean the elimination of all manual work within the manufacturing industries. There are many types of manufacturing tasks which do not readily lend themselves to automation or robotisation (Perez, 1985).

A third major area where an empirically informed, long-wave analysis leads to a very different view of the likely future sources of employment and economic activity concerns the future role of the services/information sector, relative to that of the manufacturing sector. As indicated earlier, some third-wave analysts suggest that services replace manufacturing as the major source of employment and wealth creation in the advanced industrial economies such as Britain. This is a view which is very widely held by other writers, and commentators who have addressed the issue of 'deindustrialisation'. I believe that such

perspectives are misleading in that they greatly
exaggerate the potential employment, output and
trade potential of services in the immediate future
and they downgrade that of manufacturing to a
disturbing extent. In part this arises from their
ahistorical perspective on the nature and
implications of contemporary technological changes
and from a failure to relate analysis to concrete
empirical evidence and trends.

Here it is first of all necessary to recognise
at the outset that the term services tends to be a
catch-all, residual category which needs to be
disaggregated for any meaningful analysis and that
only a proportion of the economic activities so
designated involves the production and handling of
information/knowledge (Gershuny and Miles, 1983;
Preston and Hall, 1985). Besides a large number of
services/information occupations are actually taking
place within the activities classified as
manufacturing according to the offical SIC data
(Gershuny and Miles, 1983). Secondly, however, even
if analysis is confined to the activities designated
as services within the official classifications and
data, their actual and potential contribution to
wealth creation is much less than is often assumed.
Some recent empirical evidence suggests that the
output of services industries has grown only slightly
faster than output as a whole since the early 1950s
and the evidence for long-run 'deindustrialisation'
is very slim indeed. (Bank of England, 1985; Prowse,
1985a, 1985b). Whilst the share of employment in
services has grown significantly, the share of
services in GDP has been surprisingly slow; when
changes in the price of services relative to that of
manufactures are taken into account 'it becomes clear
that the structure of developed economies has not
changed so very much since the 1950s' (Prowse, 1985b;
see Table 4.2).

Here we must also note that an important portion
of these activities is directly linked and related to
manufacturing activities. The value of intermediate
purchases of services by the manufacturing sector
rose from 13.5% of its gross output in 1954 to 18.9%
in 1979 according to one recent estimate (Bank of
England, 1985). Thus, it is difficult to conceive of
their continued existence, let alone expansion, in
the fact of a contracting or non-existent national
manufacturing sector (Perez, 1985).

Rather than viewing the services and
manufacturing sectors as alternatives or in
opposition to each other, an empirically informed

Table 4.2: Output Shares of Services in GDP and Employment (Percentages)

	GDP Value	GDP Volume	Employment
United Kingdom			
1954	44.5	47.9	42.5
1983	50.5	50.3	63.8
United States			
1953	44.5	50.4	49.5
1983	54.1	54.6	66.7
Japan			
1972	50.0	50.3	44.2
1982	53.9	48.8	52.7

Source: Bank of England Quarterly Bulletin, September 1985 (p.408)

long waves perspective suggests that there is a strong set of interlinkages between them, especially as regards the fast-growing producer services activities. This means that a robust and dynamic manufacturing sector is an essential prerequisite for the continued growth of many service/information activities as well as providing additional sources of exports earnings, employment and wealth creation in its own right.

Thirdly, it must be noted that many third wave and other commentators who downgrade the future role of the manufacturing sector tend to have a vastly exaggerated view of the actual and potential role of services in international trade. One of the defining characteristics of services is that they are purchased as they are produced and by their nature 'many services are provided directly by the producer in close proximity to the consumer' (Bank of England, 1985, p.409). For this and other reasons, such as necessary legal and regulatory obstacles to free trade, the scope for international trade growth in services is generally more limited than that for trade in goods and is largely confined to producer and financial services categories; approximately 11% of the UK's gross output of services is exported compared to about 33% of the gross output of manufacturing (ibid.); overall, services account for less than 20% of total world trade (Prowse, 1985b). Much of the focus of the services-sector enthusiasts falls on activities such as telecommunications, financial and professional services such as computer software and other advanced technical know-how and it is often argued that the advances in new information technologies will greatly enhance both the demand and possibilities for an expansion of exports in these fields. Here too it seems likely that such producer services will remain difficult to trade in the strict sense of the term because it will be more convenient to produce them on the spot rather than in a company's home territory. Indeed Prowse suggests that 'this is more true the more sophisticated the service becomes' (1985b).

Thus the most likely pattern of internationalisation of such services in the future will be a further growth of overseas investment by the larger companies and the establishment of operations and employment directly in the host countries themselves in order to gain access to local markets. Such 'foreign sales' of services may already greatly exceed the exports of services proper and they have led to a major growth in the multinationalisation of services firms in recent years (Clairmonte, 1985; Prowse 195b). This more empirically informed view of the likely developments in the services sector means that it will not of itself provide a major source of additional new employment and exports earnings in advanced economies such as Britain in the immediate future along the lines suggested by Toffler, Stonier and others.

A fourth and related point is that it is often assumed in Britain that the country has some natural or intrinsic international comparative advantage in services as opposed to manufacturing. Again this assumption does not conform to the actual empirical evidence for recent years. The data on world exports of services between 1968 and 1983 shows that the UK has lost its share in the value of world exports of services at a rate similar to that which it has lost its share of world exports of manufacturers (Table 4.3). Indeed between 1974 and 1984, the overall share of services in UK exports fell from 29% to 23% and the volume of exports of services as a whole grew less rapidly than that of manufactures (Bank of England, 1985). This decline can be largely attributed to a rapid decline in sea transport earnings and the volume of exports of other services such as air transport, financial and other services grew more rapidly than that of manufactures. Here however, it is important to stress that in recent years the decline in shipping in Britain's international trade in producer services has not been fully offset by the stability or expansion of the other categories such as air transport, tourism, banking, insurance, consultancy, telecommunications or computer software and services. For example

Table 4.3: UK Share of World Exports by Category of Trade (percentage of world exports by value)

	Transport	Travel	Other Services (a)	Total Services	Manufactures	Total Visible Trade
1968	16.0	4.8	13.9	11.9	9.6	7.2
1983	7.1	6.1	8.2	7.3	6.2	5.5

Notes: (a) Including financial services, consultancy, etc.

Source: Bank of England Quarterly Bulletin, September, 1985, p.411

financial services amounted to 3.1% of the total value of UK exports of goods and services, roughly the same as in 1974, meaning that in real terms, exports of financial services rose by only 0.1% per year over the past decade (Bank of England, 1985; Prowse 1985a). In the field of computer software and services, long assumed to be an area of strong innovative and trade performance for the UK, there has been a rapid deterioration in the trade balance in recent years (Preston, Hall and Bevan, 1985).

This empirically based analysis of the service sector's present and likely future role suggests strongly that its potential contribution to employment and wealth creation is much more restricted than many futuristic, technology-centred analyses suggest. Its likely future course of development over the coming decades will not provide the necessary employment, output or exports earnings required to compensate for a declining or non-existent manufacturing sector in Britain. Indeed, the continued growth of many services industries will be crucially dependent upon the development of a vigorous, technologically advanced and innovative indigenous manufacturing sector, rather than constituting any simple substitute for it. Whilst the share of services industries in employment creation has risen, this does not mean that their share of national wealth creation has risen or will rise in parallel in future. Only a small proportion of services is likely to be traded internationally in the future despite the advances in communications provided by the new information technologies. The notions of a 'post-manufacturing', 'third wave' or 'post-industrial' economy and the idea that services can replace manufactures in the trade balance of the advanced capitalist economies must be recognised as a chimera in the light of more concrete empirical evidence such as that indicated above. Besides, even within the manufacturing sector itself, the new IT or other high-tech producing industries will themselves provide no more than a few per cent of the sector's overall employment in the coming decades. What this means is that it can no longer be assumed by either social scientists or policy makers, that most of the advanced economies' manufacturing industries can be viewed as 'sunset' industries doomed to decline and die. Rather it is vital for such economies to renew and revive their manufacturing strength (Lorenz, 1985).

Finally we should note that this section has sought to stress the relatively limited scope for

additional employment and wealth creation within the
services/information sector in the absence of a
strong manufacturing sector. But that should not
imply that we rule out the possibility of some
expansion of service activities of a primarily
information type. Indeed it is important to recognise
that such activities have grown considerably with
each successive wave of technologies since the
industrial revolution; information and knowledge is
today an increasingly important and traded commodity
and is likely to become more so in the future given
contemporary technological and institutional innovations (Preston and Hall, 1985). The point that is
stressed in this section is that such marketed
information service activities currently make only a
small contribution to overall wealth creation and
exports earnings; thus even their rapid growth in the
coming decades does not mean that they would render a
strong and dynamic indigenous manufacturing sector
redundant in a country like Britain.

Conclusion
In this chapter I have sought to address the question
of what are the likely impacts of the contemporary
new technologies on the structure of employment and
economic activities generally in the coming decades.
To this end we first of all outlined and critically
reviewed the 'third wave' perspectives of writers
such as Toffler and Stonier. We found that this type
of contribution lacked an adequate historical
perspective on technological and economic changes,
tended towards technological determinism and gave an
unrealistic account of the likely changes on the
economic and employment in the immediate future. We
then outlined a more historically rooted and less
deterministic perspective on technological change
based on the notion of long waves, developed earlier
in this century by writers such as Kondratieff and
Schumpeter. It was argued that this served as a more
appropriate vantage point or perspective for the
analysis of the nature and implications of
technological changes. In linking this with some more
empirically based analyses we reached a number of
conclusions. The main one was that the manufacturing
sector, both the new-technology supplying industries
and more 'mature' industries, would continue to play
a major role as sources of employment and output and
exports growth in advanced economies such as Britain
in the foreseeable future. The idea that Britain's
main source of employment and wealth creation in the

immediate future would fall within the services/in-formation sector was shown to be illusory. Within manufacturing we found that the new IT supplying industries should be expected to experience some employment growth at the present time, but that within Britain they actually reduced their employment levels in recent years. This decline reflects specific economic and institutional factors associated with Britain's relative international decline as an innovative industrial power rather than simply reflecting some inherent feature of contemporary technological changes.

Notes
1. The author wishes to acknowledge support from the ESRC (formerly the SSRC) and Leverhulme Trust for grants towards research on which this contribution is based. I also wish to thank Nick Bevan for assistance with the empirical data, Peter Hall for many suggestions and guidance in the course of my research and Gillian Bogue for her superb typing and secretarial support. Responsibility for the contents and arguments of this chapter is, of course, solely that of the author.

Bibliography
Andreff, W. (1984) "The International Centralisation of Capital and the Re-Ordering of World Capitalism", Capital and Class, 22, 58-84

Bank of England (1985) 'Services in the UK Economy', Bank of England Quarterly Bulletin, September 1985, 404-14

Bell, D. (1973) The Coming of Post-industrial Society, New York, Basic Books

Blackaby, F. (ed.) (1979) Deindustrialisation, London, Heinemann

Bowlby, S. and Preston, P. (1985) 'Women and Information Technology Industries: An Historical Perspective on Technology, Gender Relations and Employment Change', Paper to IBG Annual Conference, January 1985

'British Business', 11 March 1983, 422-5

Clairmonte, F.F. (1985) 'Global Services and the TBC's, paper to CSE Conference on The Services Sector, Manchester Polytechnic, July 1985

Dosi, G. (1984) "Technological Paradigms and Technological Trajectories', in Freeman, C. (ed.) (1984)

Freeman, C., Clark, J. and Soete, L. (1982)

Unemployment and Technical Innovation, London, Pinter

Freeman, C., (ed.) (1984) Long Waves in the World Economy, London, Pinter

Gershuny, J. (1983) Social Innovation and the Division of Labour, Oxford, O.U.P.

Gershuny, J. and Miles, I.D. (1983) The New Service Economy, London, Pinter

Hall, P., Preston, P. (1985) Technological Change and The Information Industries: The Geography of Employment and Wealth Creation in Contemporary Britain, University of Reading, Dept. of Geography (mimeo)

Hobsbawm, E. (1969) Industry and Empire, London, Penguin

Kaplinsky, R. (1985) 'Electronics-Based Automation Technologies and the Onset of Systemofacture: Implications for the Third World', World Development, 13(3) 423-39

Lorenz, C. (1985) 'Why Industry Is Indispensible', Financial Times, 4 October 1985

Mandel, E. (1975) 'Late Capitalism' London, Verso

Mandel, E. (1980) Long Waves in Capitalist Development, Cambridge, C.U.P.

Mensch, G. (1979) Stalemate in Technology, London, Ballinger

NEDO (1984) Crisis Facing the UK's Information Technology Industry, London, NEDO

Perez, C. (1983) Structural Change and Assimilation of New Technologies in the Economic and Social Systems, Futures, 15(5), 357-75

Perez, C. (1985) Microelectronics, Long Waves and World Structural Change, World Development, 13, (3), 441-63

Prowse, M. (1985a) 'Services Sector Losing Share in World Markets', Financial Times, 26 September 1985

Prowse, M. (1985b) 'GATT Meeting on Services: Why Free Trade Will be an Elusive Goal', Financial Times, 27 September, 1985

Neuberg, L.G. (1975) 'A Critique of Post-Industrialist Thought', Social Praxis, 3, (1), 121-47

OECD (1981) Information Activities, Electronics and Telecommunications Technologies, Volume 1, Paris, OECD

Pollard, S. (1984) The Wasting of the British Economy, London, Croom Helm (2nd edition)

Preston, P. (1984) Innovation in Information Technology - Defining IT and Related Job and Wealth Creation Potential, University of Reading, Dept. of Geography (mimeo)

Preston, P., Hall, P., Bevan, N. (1985) Innovation in Information Technology Industries in Great Britain, University of Reading, Geographical Papers, No. 89

Radice, H. (1984) 'The National Economy: A Keynesian Myth?', Capital and Class, 22, 111-39

Rosenberg, N. and Frischtak, (1984) 'Technological Innovation and Long Waves' Cambridge Journal of Economics, 8, 7-24

Stonier, T. (1983) The Wealth of Information, London, Thames-Methuen

Toffler, A. (1980) The Third Wave, London, Pan

Toffler, A. (1983) Previews and Premises, London, Pan

van Duijn, J. (1983) The Long Wave in Economic Life, London, Allen and Unwin

Chapter Five

COMPETITION, INTERNATIONALISATION AND THE REGIONS:
THE EXAMPLE OF THE INFORMATION TECHNOLOGY PRODUCTION
INDUSTRIES IN EUROPE

Andrew Gillespie, Jeremy Howells, Howard Williams
and Alfred Thwaites

Introduction

One of the most significant developments in
industrial geography over the past decade has been an
increasing emphasis on the study of corporate
responses to competitive pressures. Moving away from
its Weberian heritage (with or without its
behavioural trappings), in which the factors
influencing the location of particular industries
(usually conceived of as consisting of single site
firms) was the object of investigation, industrial
geography has now become considerably broader in the
objects of its investigation, in which location
decisions are seen to be embedded within corporate
strategies which are attempting to respond to
increasingly severe competitive pressures. In the
most recent developments within industrial geography
(eg: Massey 1984, Sayer 1985; Storper and Walker
1983), space is being seen neither as the thing to be
explained in isolation from these broader processes,
nor as a mere appendage or residual once 'aspatial'
processes of change have been accounted for, but as
an integral and inseparable element of the industrial
restructuring that is taking place on regional,
national and international scales.

As a contentious suggestion, it can be argued
that this shift in emphasis towards a more holistic
approach to understanding industrial restructuring
in a spatial context is least in evidence with
respect to the study of the geography of high
technology industry. At least two reasons can be put
forward to explain this proposition (assuming for the
moment that it is true!): Firstly, because high
technology industry has such a distinctive pattern of
localisation, emphasis has not unnaturally focused
on trying to understand why particular locations have
'done so well'. Although questions such as 'why has

Berkshire become the leading area of high technology industry in Britain' or 'Why has Scotland managed to become a leading area of semi-conductor product-ion?', are perfectly valid questions, there is a danger - by no means insurmountable but a danger nevertheless - that answers to such questions are sought in the <u>internal</u> characteristics of such areas, divorced from - or at least not giving sufficient attention to - the external circumstances in which particular regions and the enterprises (or parts of enterprises) within them are placed. In the same way that a 'blame the victim' approach to understanding urban deprivation has diverted attention from the real causes of that deprivation, so in the high technology industry field it could be argued that an 'applaud the victor' approach runs a similar risk of missing the real point, which is presumably the nature of the competitive struggle which is being waged, rather than the characteristics of the winners or losers.

The shortcomings of the 'locality explanation' approach become most apparent when the focus of study is on small businesses, for it is here that the risk of cutting-off an area from the wider forces which to a greater or lesser extent will have shaped it is greatest. In the UK context, the most extreme example of this trap is evident in some of the attempts to explain the success of the central lowlands of Scotland in high technology industry terms. By focusing on new firm formation as the object of investigation (for example, in Oakey, 1984), the point can all too easily be missed that the 'success' of this region is almost entirely due to the European investment location decisions of a comparatively small group of (predominantly US) transnational corporations.

A second reason which can be put forward to explain the somewhat 'traditional' type of approach to studying the geography of high technology industry concerns the relative failure of those of a radical persuasion - who would have been expected to have adopted a more holistic perspective - to have confronted adequately the nature of the restructur-ing processes taking place within these industries. Sayer (1985, 1986) suggests that this is partially because such industries have not 'fitted in' well to the main currents of the post-Braverman debate in the radical literature concerning issues such as de-skilling and Fordist and neo-Fordist labour processes. More fundamentally,

there are also some rather restricted views of
competition in the radical literature ... there
is a tendency to see competition only in terms
of price competition for a largely fixed stock
of goods, thereby underestimating the
importance of product innovation and technological complementarities in competition. (Sayer,
1986, p.108).

With respect to high technology industry (or
more specifically to electronics), this tendency has
resulted in undue emphasis being given to strategies
which have attempted to reduce costs by seeking out
cheap labour cost locations for assembly operations
(exemplified by the new international division of
labour approach of Frobel et al, 1980), and
insufficient emphasis being given to the key
competitive dynamics of these industries which is
product innovation (Freeman, 1974; Porter, 1980).

Our aim in this paper is to attempt to provide a
bridge between the 'location specific' and the
'international restructuring' types of approach to
studying the geography of high technology industry.
The approach we adopt is to begin by examining some
of the dominant types of corporate response to
competitive pressure in one sectoral and
geographical sub-set of 'high technology' industry –
that involved in the production of information
technology in Europe – and then to consider some of
the regional development implications of these
dominant forms of restructuring process.

Methodology
The paper draws upon some of the interim findings of
a project concerned with the locational dynamics of
Europe's IT industry which was funded by the
Commission of the European Communities. The primary
method of investigation was through structured
interviews with senior management in 147 companies.
Initially, within each of four chosen product sectors
(data processing hardware, telecommunications
equipment manufacturing, software engineering and
active components), major multinational operators
(including those from Japan and the USA) and
nationally-oriented companies were selected for
interview (a list of the major companies interviewed
is provided in Table 5.1). Secondly, a number of
regional case studies, which involved interviews
with samples of companies, including SMEs, operating
in the four sub-sectors as well as with relevant

115

institutions, were undertaken in various parts of Europe's 'less favoured' periphery (these being Berkshire and Strathclyde/Lothian in the UK, the East and West regions of Ireland, the Languedoc and Moselle-Lorraine regions of France, the Lombardia, Naples and Abruzzi regions of Italy and the Athens and Salonika regions of Greece.

Table 5.1 Multinational and National Firms Interviewed

1.	Siemens	2.	Philips
3.	STC	4.	ITT
5.	IBM	6.	Honeywell
7.	Plessey	8.	Olivetti
9.	Telettra	10.	Italtel
11.	DEC	12.	Hewlett Packard
13.	National Semiconductors	14.	Thorn-EMI
15.	Wang	16.	Nixdorf
17.	Ferranti	18.	Machines Bull
19.	CIT Alcatel	20.	TRT
21.	Northern Telecom	22.	NEC
23.	Amdahl	24.	Fujitsu
25.	Gould	26.	Prime
27.	SCI	28.	GSI
29.	CAP GEMINI SOGETI	30.	SGII
31.	CISI	32.	Logica
33.	CAP	34.	Data Management
35.	Sipe Optimatia	36.	Datamont
37.	Thomson Semiconductors		

The 'aide memoire' which was used to structure all the corporate/enterprise interviews was primarily concerned with the following issues:

-   the process of resource allocation decisions within the company and how this process has changed and may change;
-   the interactions, and how they are changing, between the product range, major market and customer segments and geographic markets;
-   the input/output processes of the company and specifically the role played by manufacturing and how this has changed and may be expected to change;
-   the processes by which the company expands its markets and the implications of this upon company structure and organisation;
-   the processes by which the company develops its product range and the implications of this upon company structure and organisation;

- the spatial development and trends of company
  structure and organisation and factors
  affecting this development.

The responses of Europe's Information Technology industries to increasing international competition

It was suggested above that product innovation is the
key competitive strategy which characterises the
information technology industries, reflected in a
high level of commitment within the industry to R and
D. Increasingly, however, product and process
changes are becoming coincidental and complementary
(Porter 1983), while the innovation 'tread mill'
(Freeman, 1974) evident in the electronics sector is
being reflected in shorter and shorter product life ←
cycles (Goldman, 1982) which, in turn, has profound
implications for the way markets are perceived and
served. The perceived period of the product life
cycle in which profits can reasonably be earned is
certainly no more than three years and often as
little as two. Thus, the **volumes** needed to cover the
design, development, production and marketing costs
and to sustain the profitability of the company have
grown enormously. In consequence, there has been
considerable investment in process automation which
has produced the economies of scale necessary to
operate internationally, and an increasing tendency
for the major corporations to introduce global
products, undifferentiated between nations and
regions (with the possible exception of the
telecommunications sector in which the PTTs still
have sufficient monopsonistic power to fragment the
market along national lines). In the following sub-
sections we consider each of these inter-linked
developments in turn, from the perspective of the
competitive pressures upon Europe's IT producers.

That these pressures are growing there can be
little doubt, reflected in an increasing domination
of world IT markets by US and Japanese producers. In
their report to the European Commission, McKinsey and
Company (1983) attempted to document the global
position of the European IT industry. They entitled
the report "A Call for Action" which reflected their
analysis of the situation and the disparity that
exists between the European and the USA and Japanese
industries and the need for immediate concerted and
collaborative **European** action if parity was to be
achieved by the year 2000.

Product innovation, R and D and the competitive position of European IT producers. In all the sectors studied the importance of fundamental research into the enabling scientific technologies was seen as a pre-requisite to a thriving IT sector. However, it was evident that actual costs involved in such activities were becoming prohibitive. These costs, however, have to be absorbed if a presence is to be maintained at the leading edge of the technology. There appears to be an <u>absolute</u> level of commitment, in terms of R and D expenditure, necessary to maintain leading edge presence in certain IT sectors. In telecommunications equipment, for example, it has been estimated that the cost of developing a digital switch ranges from $700 million to $1 billion. At the 'normal' (for this industry) 7% of sales devoted to research, the sales required to cover the development costs are almost $14 billion, which as McKinsey and Co (1983) point out, is well above the size of the largest European national market. The fragmentation of the European market along national lines is reflected in Europe's IT producers being small in international terms (only Siemens and Philips appear in the world's top ten IT producers in terms of turnover). Significantly, only these two companies have been large enough to approach the R and D investment levels required to maintain a competitive toehold in the most competitive of all IT markets, that for standardised integrated circuits (Dosi, 1981).

In all the sectors studied changes are occurring in the fundamental technologies and consequently numerous product possibilities are emerging. For example, in data processing the convergence of data processing and telecommunication technologies and the development of optical processors is creating new opportunities and posing major problems. But it is perhaps in semiconductors that most dramatic effects in technological change can be seen, for example, in the development in technologies required to achieve three dimensional sculpturing of silicon wafers and the printing of sub-micron lines in order to construct 4 megabit d-rams (dynamic random access memories).

The need to commit substantial resources to R and D stems from the fact that the aggregate level of technical change within the industry renders products technically, and thus often commercially, obsolete after a limited period. Once this has happened the level of prices collapse and the viability of individual companies is threatened. For

example, the 256K d-ram market in the US collapsed to
$1.2bn in 1985 from $3.5bn in 1984, within some
individual cases the prices of 256K d-rams falling by
over 80% in 1984. A heavy commitment to R and D and
innovation is the only route to maintain premium,
profitable pricing. However, the risks are high and
this has forced collaboration between different
manufacturers. In the semiconductor sector the only
remaining major European force in the development of
4 megabit d-rams is a joint venture between Siemens
and Philips. Although each company has committed
substantial funds to the project and these sums have
been matched by an almost equally large subsidy from
the Dutch and West German governments, the project is
behind that of its major competitors and there has
already been a reduction in expected prices by a
factor of 7. The original intention, however, was to
develop an in-house 1 megabit d-ram, but this has
been dropped and licences have been negotiated with
Toshiba.

The technological leadership in all IT sectors
is undoubtedly held by the US and Japan and there is
intense competition between these two nations. One of
the major explanations for this technological
dominance is the use of industrial targeting by the
Japanese and American governments. Industrial
targeting is defined by US Department of Commerce as

> the selective use of instruments by a government
> in order to enhance the competitiveness of a
> particular industrial sector - it is a
> concentrated form of industrial policy. The
> instruments can be combinations of fiscal and
> protective trade measures applied in varying
> degrees of intensity. In a number of instances
> governments also create or permit the formation
> of industry groupings which intensify their
> industry's concentration in order to achieve
> larger scale economies, reduce duplicative
> activities, or influence the direction of
> industry development. These groupings can
> simply be research co-operatives made up of
> major industry members and permitted by
> exceptions in anti-trust laws or they can be the
> nationalisation of the entire industry.

In the words of the US International Trade
Commission, industrial targeting is 'co-ordinated
actions that direct productive resources to give
domestic producers in selected industries a
competitive advantage'.

The effect of industrial targeting is to create a virtuous cycle in which the front end cost of new products and technology including the commissioning of actual production facilities is borne by the government. In the US the major government agent involved in industrial targeting is the Department of Defence and in Japan it is the Ministry of International Trade and Industry. Thus the Americans tend to use defence procurement and the Japanese targeted commercial application as the catalyst.

It is within the context of the technological dominance of Japanese and American companies in the IT Sector and the policy framework of industrial targeting that the work of the EEC's Information Technology and Telecommunication Task Force (ITTTF) and in particular the ESPRIT and RACE programmes should be interpreted. ESPRIT (European Strategic Programme of Research in Information Technology) is a 5 year, $1.5bn programme which is primarily concerned with promoting pre-competitive research in advanced micro-electronics, software engineering, advanced information processing, office systems and computer integrated manufacturing. Each project draws together researchers from different organisations and countries in collaborative ventures of which 50% of total costs are borne by the EC. The ESPRIT interim report (European Commission 1985) has concluded that 'despite initial scepticism on the part of many IT companies there is now unanimous agreement as to the initial success of ESPRIT particularly with respect to the promotion of trans-European co-operation between IT organisations'. The review revealed evidence that pre-competitive collaboration at the R and D stage had, in some cases, been extended into other joint ventures (e.g. the Bull, ICL and Siemens Software Centre). Also of importance was the evidence that ESPRIT has not only resulted in projects that were broader in their scope and more ambitious than originally intended, but that these projects were being rapidly progressed by the larger resources dedicated to the research.

The defensive strategies which are emerging at the European level are consequently to try and narrow the 'technological gap' between European producers and their US and Japanese competitors which had become so marked in the 1970s. Even if the strategy fails, the only viable alternative strategy, of accepting the inevitability of the gap and institutionalising it via technology and product licences negotiated from USA or Japanese companies, carries severe dangers. McKinsey (1983) argued that

120

rather than restoring some sort of competitive equilibrium, technological dependence would be symptomatic of a 'vicious circle of decline', in which the reliance of domestic producers upon imported technology and products would create an embedded product/customer base which the originator of the product/technology could subsequently capture.

The implications of the 'globalisation' of markets for Europe's IT producers. The internationalisation of markets for IT products has preceded the internationalisation of production per se, in accordance with Palloix (1975), who predicted a sequence in which the internationalisation of commodity capital (i.e. trade) takes place earlier than that for either money capital (portfolio investment) or productive capital (foreign direct investments). The competitive drive to maintain profitability over very short product life cycles by achieving near global reach has had profound implications for Europe's IT producers.

Three distinct trends are emerging. Firstly, within the context of Europe's fragmented national markets the concept of the 'national champion' is becoming obsolete. It is now almost prohibitively expensive for a European nation to support, by itself, a national IT champion which is vertically integrated from the component level through to final product. The only possible exception to this is within the software sector, but the evidence illustrates the increasing importance of multinat-ional software houses. For example, US companies feature prominently in the national lists of top software houses produced for the European Computer Services Association and this is reflected in the aggregate European figures; US software houses accounted for almost 27% of computer service revenues in 1983 and for 17% of the independent packaged software market.

The need to develop international forward (market) and backward linkages is reflected in the numerous inter-corporate agreements that have been restructuring the global IT production sector. These changes are perhaps best illustrated in the telecommunications equipment sector as manufacturers need to exploit the opportunities created by digital technology, the rapid growth in telecommunication use and deregulation, particularly in the US and to a lesser extent in the UK. Van Tulder and Junne (1984)

121

have documented many of the links between multinationals in the telecommunications sectors. There has, however, been a continuing dynamic that has resulted in further concentration in the industry. Within the UK, STC has essentially withdrawn from the digital exchange equipment market apart from its residual interest in selling ITT's System 12, while GEC has bid for Plessey largely to rationalise its telecommunications activities centred upon System X. Within France the telecommunications equipment industry has basically been concentrated in CGE although it is looking for further collaboration in an international context with AT and T. In Italy STET and Fiat have agreed to merge their telecommunication interests (respectively Italtel and Telettra) and to seek an international partner (AT and T?). Siemens of West Germany has established a joint venture with GTE in the US which gives Siemens the best access yet achieved by a European company to the lucrative US market. Similar developments can be seen in both the active component and data processing hardware sector. A tentative description of corporate linkages (shareholdings and technological agreements) between some European IT manufacturers and other major corporations is illustrated in Figure 5.1.

Thus, it is evident that there is a major restructuring taking place of the IT production sector at the international level and it will only be those consortia that achieve the internationalisation of their products that will remain in the top league. Those companies that now have substantial export businesses that are unable to secure advantages in the current realignment of the industry will retreat into their national markets or perhaps into peripheral but related markets (e.g. Honeywell into control systems).

Secondly, on a national scale major companies will remain but will become, in effect, system integrators. These companies will create a strong national market presence based on a wide product range through a combination of their own design and manufacturing expertise and tied OEM (original equipment manufacturer) agreements coupled with their own label sales. To the consumer the diversity of product sourcing will remain transparent. There are already a number of examples of previously major international IT companies being forced back into their national markets and becoming reliant upon key assemblies and products from third party sources. In

Figure 5.1: Tentative Description of Some IT Corporate Linkages

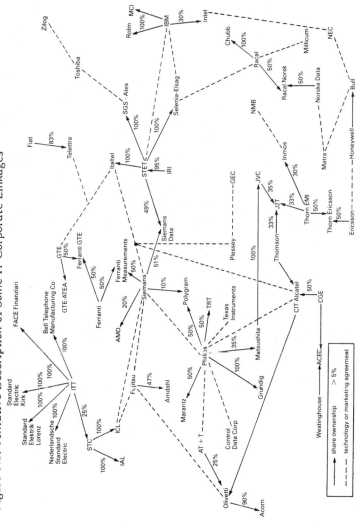

the data processing field ICL is a classic example,
its large mainframe computers are largely dependent
upon Fujitsu technology and in the super micro/mini
market it sells a US product (ICL PERQ).
The logical development of the system
integration route is for the company to become little
more than a market responsive selling vehicle and
only retaining the technical expertise needed to cope
with the idiosyncratic nature of national standards
and protocols. This process adheres to the 'vicious
cycle' of decline identified by McKinsey and
described above.
Thirdly, an inevitable consequence of the
internationalisation of IT products and the focus of
major producers upon major markets is the creation of
niche market opportunities for smaller companies.
However, these niche markets are by definition
limited in their size and this can pose major
problems for the expansion of individual firms. In
the software sector the importance of niche markets
in which a small firm is able to concentrate upon a
specific application package is of fundamental
importance to new firm formation and survival. This
was evident from our research where numerous software
houses interviewed were only active in servicing
specific markets, generally on a national scale. In
the active component sector the oscillation between
custom-chip design and merchant chips by the major
semiconductor manufacturers and the rapid growth in
the use of semiconductors in consumer products (e.g.
washing machines) has led to niche market
opportunities especially for semiconductor and
integrated circuit design houses. The research
identified several such companies. Another niche
market created in the active component sector is the
design of novel but dedicated circuits for specific
applications using standard merchant components. The
front end design costs for custom semiconductors are
generally in excess of £40k (the prices quoted are
typical industry figures quoted in interviews
(Summer 1985) and this necessitates high volume
production to achieve unit prices measured in pence.
Many companies have a requirement for custom
semiconductors but only need low volumes and it is
these companies who have created the market for the
design of novel circuits. Given that these circuits
often cost £2/3,000 to design, the savings on low
volumes are obvious.
One of the major structural weaknesses in IT
markets is the absence of any coherent set of
international standards governing, among other

things, the protocols of product compatibility and interconnection. The lack of industry-based standards results in standards becoming proprietorial, and de facto controlled by those companies with the largest market share. In fact proprietorial standards are used to their maximum advantage in the marketing strategies of the major companies. The ability to create uncertainty and thereby inhibit potential customers' purchases of competitive products is only one way in which major companies are able to exploit proprietorial standards. But, perhaps more significantly, proprietorial standards are used to fundamentally influence the development of peripheral but related markets. The use of standards in this way is clearly illustrated by IBM. The development of the personal computer market was initiated independently of IBM who were 'late' entrants yet because of their dominant embedded customer base in the mainframe sector IBM were able to impose their standards on the whole industry. The growth of IBM PC sales and the proliferation of IBM PC application software gave competitors a stark choice, produce IBM compatible products or risk your business.

A similar scenario is developing at the moment in terms of local and wide area networks. The lack of international open industry standards is particularly a problem for small to medium sized enterprises (SMEs). Often the basis upon which these firms are able to build a competitive advantage is through technically advanced solutions in new markets. Yet by definition these firms are not large enough to be able to create proprietorial standards to their advantage and are thus exposed to the competitive threat posed by 'late entrants' who are major IT companies. For example, we interviewed a SME in the UK that claimed to have 25% of the PC local area networks (LANS) market yet were facing considerable problems, because IBM had created uncertainty in the market by its ambiguous statements over its own intentions about LANS.

Internationalisation and the changing organisation of production systems. Not only are companies having to <u>sell</u> in global markets, but <u>production</u> itself is increasingly becoming organised in accordance with market-serving strategies.

It was evident from an analysis of the corporate structures of the major IT companies that there has been a decline in the national production centres manufacturing a wide range of products and a growth of territorial (often inter-continental and always transnational) production sites dedicated to a single product or limited product range. For example, the IBM plant at Greenock in Scotland has responsibility for the production and distribution of the IBM PC throughout Western Europe, the Near East and Africa, although originally (1951) it was responsible for the manufacture of all IBM products sold in the UK. In parallel with this development of product specific plants has come the affirmation of control from corporate HQ manufacturing director-ates. Although offshore (to the corporate HQ) facilities are usually administratively part of the corporate legal entity in the country in which they reside this is not translated into real local management control in terms of resource and product allocation decisions. Ironically though the country operations are generally the source of capital to fund the manufacturing activity (e.g. Hewlett Packard).

As well as affecting the organisation of production within companies, the internationalisat-ion of markets and their increasing volatility is also beginning to affect input linkages - i.e. manufacturing systems more generally defined as combinations of inter-linked enterprises.

The fundamental pressure upon manufacturing systems is the desire of companies to become more responsive to the needs of the market. In consequence firms are implementing production strategies that both increase the flexibility of their operations and reduce their costs. Thus there has been a rapid move, but only over the last two to three years, to Just-in-Time (JIT) away from Just-in-Case (JIC) manufacturing systems.

Alternative names for JIT are kan-ban (Japanese) or continuous flow manufacturing. JIT is a method of production organisation that allows the delivery of product to the customer directly from the production line (via the distribution network) and not from a stock of pre-made (anticipated orders) products. Thus the rate of production is determined by the current level of sales and in doing so the production side of the firm becomes subservient to the market and this forces organisational and operational change.

Perhaps the most obvious effect of JIT is that because the consumption rate of inputs becomes increasingly difficult to predict and holding stock is expensive and risky (especially given the volatility of component product prices), firms have minimised their own stock levels and placed the emphasis upon their suppliers to deliver 'Just-in-Time'. In consequence it has often been argued that JIT is essentially a stock holding externalisation process with profound consequences upon the location of suppliers, i.e. - they must be in close proximity to main manufacturing facilities (e.g. Estall, 1985). It was evident, however, from our research that the move to JIT reflects far more fundamental issues which are concerned with the international-isation of IT markets than merely the externalisation of stock holding.

Concomitant with the introduction of JIT, companies redefine their in-house manufacturing and reallocate resources only to those activities which are perceived to be the major sources of value added and capable of growth. Activities which fall outside these criteria are externalised often by spinning them off into independent firms. This analysis of value added is extended to each potential supplier so that each enterprise in the system is responsible for that element in which it can achieve greatest value added and thereby a process of aggregating competitive advantage is formalised into the total production system. These decisions are made on an international scale for each product has to be internationally competitive. Thus one of the important criteria used for screening potential suppliers is their ability to operate effectively in an international market. The concept of the 'hollow corporation' has developed in recent years as the full implications and possibilities of JIT are explored. In essence the 'hollow corporation' describes a phenomenon whereby the design and all conceptual work related to the product is contained in-house but all actual processes are sub-contracted. For this system to work all suppliers have to be connected via a sophisticated communications network.

In exchange for a commitment to support the JIT manufacturing systems at the customer's plant, the supplier is able to reduce some of the risks of being in business. Not only does the main manufacturer place volume orders at economic prices but it also provides a substantial input into all aspects of the business, including product development, process technology (often including the machine tools) and management development. Thus the enterprises within the JIT system become mutually supportive and capable of being responsive to the needs of a volatile market. However, the real control in the system remains with the final product manufacturer and in essence JIT is a system that gives these manufacturers 'arm distance' but almost total control over their suppliers.

The operation of JIT varied between the different sectors studied and ranged from being almost non-existent in the software sector, largely due to the 'artisan' production process, to being widespread in the active component sector. Many of the active component manufacturers have fully implemented JIT. In some cases the stock held of consumables is down to just several <u>hours</u> production (it is measured in hours not volumes), with the plant dependent on twice daily deliveries. In other cases, particularly for 'tools', no stock is held at all.

Inevitably to maintain the production process meticulous attention must be paid to quality; in fact the system cannot operate unless suppliers have the capability to 'ship to stock'. This means there is no inspection of incoming material. In the semiconductor industry the target is fault free product; already it is becoming common within the industry to measure quality in terms of parts per billion. These expectations on the quality of output are reflected throughout the system and failure to meet delivery schedules (both on time and quality) result in <u>no</u> business not reduced business. If this rejection does occur then the re-entry problems faced by a firm are substantial, if not insurmountable.

From the interviews it emerged that the trend to JIT is accelerating and becoming more widespread in all the sectors studied. In fact in the data processing sector the trend that emerged was one of whole sub-assemblies, e.g. VDU, keyboards, being sub-contracted and delivered directly to the final customer independently of the 'label' manufacturers.

Process automation and changing skill requirements.
Considerable investment in process automation has produced the economies of scale necessary to operate internally. The achievement of scale economies via capital intensity, in turn associated with yields and quality control, is particularly important in the production of semi-conductors. As a result of success in achieving these scale economies, the proportion of labour costs to total costs has fallen, and low labour costs locations have become less attractive for semi-conductor production. Partly this effect is also due to more complex product, in which labour

costs in the assembly stage constitute a comparatively small proportion of overall costs. One US estimate suggests that labour costs for assembly operations were 33% of total costs for simple integrated circuits, but only 4% for complex integrated circuits (UNIDO, 1981). Further, the assembly operators are themselves becoming increasingly automated (Rada, 1982), reducing the attraction of cheap labour locations. As a result of these developments, some semiconductor production which had gone offshore is being re-introduced back into Europe. For example, we interviewed a major UK company which has recently opened a highly automated semiconductor plant in Berkshire, replacing its former overseas low labour cost sites, including the Philippines. It should nevertheless be stressed that offshore production by European companies has anyway never been as important as it has been for the US merchant chip makers, so any reduction in offshore activities will not make appreciable differences to European production levels.

That the skill structures of the IT industries depart substantially from those which prevail in most other manufacturing industries has for long been recognised. The labour requirements tend to be polarised between a large segment, and a relatively growing segment, of highly qualified labour (scientists, technicians, managers) and an unskilled, semi-skilled workforce engaged in assembly. This trend towards bifurcated workforce has resulted in skilled <u>manual</u> labour becoming very much under-represented in the IT production sector.

The concept of a bifurcated work force, however, disguises the extent to which the highly qualified scientific and technical staff are becoming increasingly integrated into the production process. In the equipment manufacturing sectors the heavy investment in automation and the need to achieve extremely high standards of quality has increased the direct employment of professionally skilled employees on the production line. Additionally, the changing sources of value-added, for example in telecommunications, has resulted in the production environment moving to the scientists and technologists (e.g. software development).

In the more general context of 'labour markets' these changes in the skill profile of employees have profound influences upon the ability of companies to operate effectively in certain types of labour markets. In the past the requirements for semi-skilled and unskilled <u>manual</u> labour were an overriding consideration and the availability and quality of such labour influenced the performance of production establishments. The reverse is now true and it is the availability of highly educated and highly skilled labour that is becoming an increasingly important issue for the IT production sector.

For example, in the software sector it can be argued that the skill profile only includes those who are highly educated and professionally trained. As such the sector represents the 'end point' of the trends discussed above. It was evident from many of the software houses interviewed that only graduate qualified individuals were recruited. One of the possible explanations of this need for such highly skilled people is in the nature of the production process itself. Although considerable effort has been, and continues to be devoted to software production tools it is still difficult to disembody the various stages of the development and delivery of a software product. Thus the development of a software product is the equivalent to producing a piece of customer specified R and D. It can therefore be hypothesised that as production tools become available the production process itself will become divisible and create opportunities for a wider range of skills to be effectively employed. However, evidence from the software production tool research programmes suggests that the opportunities for the mass employment of programmers and code writers are limited as it is these functions which will be automated.

In the semiconductor/active component sector, employees are the major source of 'containments' that reduce the yield of the production process. There is, therefore, a trend to automate the processes and this reduces the demand for employees overall, but particularly for semi-skilled employees. Also the increasing degree of automation necessitates the employment of highly skilled and educated employees which adds a further pressure upon the employment opportunities for semi-skilled workers.

Over and beyond the skill profile of the workforce and individuals the industry has adopted another set of criteria that can best be described as "attributes". These attributes which essentially focus upon adaptability, flexibility and commercial awareness of an individual are seen as being equally if not more important than skills, traditionally defined. The importance of these attributes reflects the need for many of the reasons discussed above of

IT companies to enhance their flexibility and responsiveness. Therefore IT companies have generally adopted thorough recruitment processes in which it is explicitly recognised that the possession of skills is no longer a necessary and sufficient condition to achieving employment. For example, it is not uncommon for applicants to be subjected to extensive psychological initiative and character tests before any assessment of their 'skills'.

Further the major IT companies interviewed have adopted 'progressive' employment policies again designed to ensure the flexibility of the organisation and maintain and enhance employees' job satisfaction. This latter point is significant, for within many segments of the IT sector people are the key resource even in manufacturing activities, and compared to other sectors, the movement of employees between firms is accepted as an industry norm. However, the loss of key staff can be damaging and in competitive labour markets the replacement of staff is at its minimum time consuming but is also often expensive.

It was evident from the interviews that the employment practices used by IT firms in all sectors and almost regardless of company size were diverging significantly from practices in other industries. The corner-stone of these new employment practices is the individualisation employer/employee relations which has profound consequences upon the whole set of relationships and practices within the firm. For the individual the new employment practices mean that there are no salary scales and grades. Thus the basis of pay is upon an individual's contribution to the firm and not related to the output of that person. One of the main advantages of this method of payment is that the flexibility of employees is increased for the perceptions and pay are not related to a specific task. The process, however, necessitates a formalised staff appraisal system that is seen to be open, explicit and fair.

Trade Unions are generally not recognised in IT firms. The reasons for this are complex but individualisation removes one of the basic roles of trade unions, that of pay negotiation. Additionally, the role usually played by trade unions in discussing with management and communicating corporate decisions and plans to employees has been supplanted by the extensive use of direct employer/employee communications, e.g. briefing groups. These direct communication methods are genuinely two-way and not 'management tell'.

Thus, through the process of the individualisation of relationships and direct communications, employers have attempted to increase job satisfaction and the flexibility of the enterprise. In so doing, however, these firms have successfully eclipsed the role of traditional trade unions and in general such unions are not recognised. One of the best examples of these new employment practices in Texas Instruments where they have managed to achieve the individualisation of all employee/employer relationships in the UK.

It appears that it is the major multinational companies (US and Japanese) which are setting the pace in developing new employment practices, particularly those firms whose total activities are dedicated to the IT sector. This has posed problems for many European companies which have developed their IT activities from a base in electro-mechanical engineering. It was evident that expectations of key employees in IT differ substantially from what is the norm in other sectors. This dichotomy is perhaps one of the explanations for the 'skill shortage' problem, for many of the 'pure' IT firms interviewed argued that their best recruitment grounds were the traditional electro-mechanical companies that have moved into electronics. The arguments were that employees in such firms were often stifled, given limited responsibility, were poorly motivated and caught up in internal politics and bureaucracy and thus keen to move to companies with progressive employment practices.

Implications for regional development prospects in Europe

The internationalisation trends outlined in the previous section clearly have implications not only for European IT production, as a whole, but also for production within Europe's constituted member states and near regions. The European IT industries already display marked geographical concentration, with the major metropolitan regions of France, the Low Countries, Germany, Northern Italy and the UK dominating the geography of IT production in Europe (Gillespie et al, 1984; CURDS, 1986). In this section we attempt to summarise, very much in outline terms, some of the implications of the trends we have described earlier for the prospects for IT production within Europe's less favoured regions.

1. **Product innovation**: Given what is already known about the geography of R and D activities (e.g. Malecki, 1980 and 1986; Buswell, 1983; Savey, 1983; Howells, 1984), of product life cycles (e.g. Norton and Rees, (1979) and of product innovation (e.g. Twaites et al, 1981; Oakey et al, 1980 and 1982; Goddard et al, 1986), there can be little doubt that the dominant response of IT producers to increasing competition - i.e. high rates of product innovation resulting in shorter product life cycles - will enhance the relative prospects of the existing centres of IT production in Europe, particularly those with a strong presence of R and D activities. For regions with limited presence in R and D, the truncated product life cycles which are becoming the norm will pose threats for their existing IT production capacity (as the UK's peripheral regions have found in recent years with respect to telecommunications equipment manufacturing, as their electro-mechanical product ranges are being replaced by electronic equipment manufactured in other regions). Nor can it be assumed that the presence of research activities within Higher Education Institutions (HEIs) within less favoured regions will be able to compensate for the lack of in-house R and D in a region's enterprises; we found very little evidence in our survey of technology transfer between European HEI's and the IT companies. In Europe at least, R and D is primarily carried out within in-house company facilities, and what few links with HEIs do exist seem quite unconstrained by geographical proximity.

Given the scale of R and D commitment necessary to compete head-on in world markets, the European Community policy objective of trying to strengthen the European NIT industries by fostering trans-national research links and by trying to create a sufficiently integrated European market to enable economies of scale to be reaped would seem to have a lot to recommend it. However, while this strategy makes sense at the level of the NIT industry within Europe as a whole, its impact may be less positive for Europe's less favoured regions and for those member states with relatively weak IT industries. For example, both ESPRIT and RACE have focused the Community's R and D efforts into major corporations. The dominance of the major 12 European producers in ESPRIT (these being Bull, CGE, Thompson (Fr); AEG, Nixdorf, Siemens (FDR)); GEC, ICL (now owned by STC), Plessey (UK); Olivetti, STET (It); Philips (NI) is a reflection of the policy objectives which themselves largely stem from the McKinsey analysis of the strengths and weaknesses of European IT companies. In the implementation of the first phase of ESPRIT the ITT task force were mindful of such criticisms and attempted to ensure a wide spread of Community participation; nevertheless, as Figure 5.2 indicates, the location of R and D establishments (Universities and detached corporate R and D units) participating in ESPRIT reveals limited involvement by less favoured regions.

2. **Market internationalisation:** Although in one sense the increasing internationalisation of markets provides a competitive stimulus to the largest corporations with the resources and expertise to market their goods internationally and to reap economies of scale, as suggested earlier this development also provides new opportunities for smaller companies - and, by implication, for IT production in some of the non-core regions - to identify and exploit niche markets which are arising as the largest companies concentrate on global product reaching their main markets. We return to this possibility in the final section.

3. **Changing organisation of production systems:** The growing internationalisation of production systems, in the wake of the internationalisation of markets, poses a number of threats for less favoured and/or dependent regions but also, potentially at least, some opportunities as well. One of the most predictable threats will be the reduction in existing levels of 'national' autonomy for subsidiaries of TNCs as production is increasingly organised to serve large geographical markets (transnational and often transcontinental) with single products rather than serving national markets with broad product ranges.
Important implications can also be foreseen associated with the rapid spread of Just-in-Time manufacturing systems in the IT production industries. Rather than JIT dictating the geographical proximity of suppliers and customers (as some have perhaps mainly seen it), JIT systems are in fact part of a strategy of implementing global sourcing. The key to successful participation in a JIT system is concerned more with the ability to meet the most exacting quality control standards rather than the geographical proximity of supplier to customer. Existing component suppliers will come

Figure 5.2: R&D Establishments Participating in ESPRIT

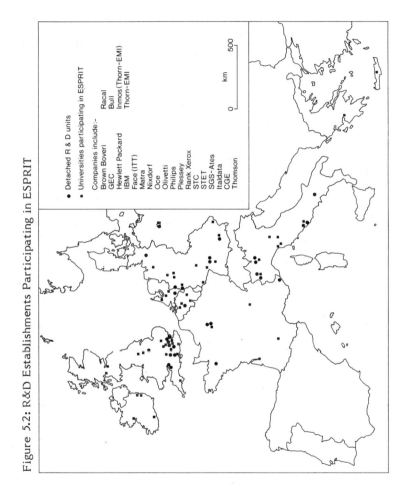

- Detached R & D units
- Universities participating in ESPRIT

Companies include:-
Brown Boveri
GEC
Hewlett Packard
IBM
Face (ITT)
Matra
Nixdorf
Oce
Olivetti
Philips
Plessey
Rank Xerox
STC
STET
SGS-Ates
Italdata
CGE
Thomson

Racal
Bull
Inmos (Thorn-EMI)
Thorn-EMI

under increasing pressure to raise these standards -
often requiring major investment in process
equipment and in human capital - if they are to
continue to compete in markets organised along global
lines. At the same time, however, the strategy of the
high value added 'hollow corporation' is likely to
increase the range of inputs which are sub-contracted
out, perhaps providing new opportunities for smaller
companies (but only if virtually complete dependency
on the core corporation is accepted, given the way
JIT systems 'lock in' component suppliers).

4. Process automation and labour requirements: The
starting point for any consideration of regional
development prospects in relation to IT is that
'jobless growth' (Morgan and Sayer, 1983) is very
much the industry expectation. Thus it now seems
clear that whatever benefits will accrue to regions
which are 'successful' in maintaining or even
increasing their presence in the IT production
industries, the creation of net new jobs is unlikely
to be one of them. Indeed, in particular sub-sectors
- most notably in telecommunications equipment -
further major reductions in employment are foreseen,
arising from a combination of product and process
changes within a market context of chronic over-
capacity.
    With respect to labour availability as a
locational attraction, the implications of our
earlier discussion point clearly to the reduced
importance of semi and unskilled labour reserves for
IT production. Regions which have come to occupy a
particular position in the spatial division of labour
within IT because of the existence of such reserves
are unlikely to benefit from branch production
investments to anything like the extent they have in
the past. Labour is still an important locational
resource, however, but only in the sense of the most
highly skilled and highly educated elements of the
labour force. It is actually in the sense of
providers of this type of labour that the HEI's in
less favoured regions can be considered important,
(rather than as centres of R and D and technology
transfer). Certainly the relative success of
Scotland and Ireland is attracting mobile IT
investment projects would seem to be related to their
perceived high quality output and availability of
graduates in the relevant technological fields.
    Finally, the notion of 'skill' as a social
attribute related to with a worker's attitude and

flexibility as much as with his/her qualifications or
technical proficiency may have implications for
regions which are perceived to have particularly outmoded (in the eyes of the IT companies) forms of
worker organisation and industrial relations
climate. However, the experience of companies such as
IBM in Greenock, and those of American companies more
generally in Scotland and South Wales, would not
suggest that the regions of strong traditional labour
organisation necessarily inhibit the imposition of
rather different industrial relations regimes
(probably because of the gender and age
characteristics of the workforces recruited being
different in the IT industries from the regional
norms).

Towards an IT production strategy for Europe's less favoured regions

In the final section we outline some of the issues
which would need to be addressed in the formulation
of policies designed to encourage or foster IT
production in Europe's less favoured regions, within
the context of the competitive pressures operating
increasingly at an international scale.

For a less favoured region (or country in the
European context) with limited existing IT presence,
at least three kinds of strategy can be envisaged for
promotional regional IT production:

i   To attempt to attract trans-national corporations to set production facilities.
ii   To attempt to build up 'national champion'
companies to compete with the TNCs,
particularly in the supply of home markets.
iii To attempt to identify 'niche markets' at
national and international scales, which could
be exploited by indigenous companies, given the
existence of targeted industrial policies.

i **TNC inward investment strategy**: Because of the
rapid growth of the IT industries, there have been a
number of major investment decisions by US and
Japanese companies from which some of Europe's less
favoured regions have benefited. Ireland in
particular has pursued for over a decade a policy of
attracting IT production facilities, while Scotland
has become Europe's leading area of semiconductor
production, almost entirely as a result of inward
investment from the US (there are, for example, six

major foreign-owned wafer fabrication plants located
in Scotland). The Scottish Development Agency
(founded in 1975) identified electronics as a major
growth sector which Scotland could secure a share in,
and established an Electronics Division to develop
and pursue a strategy to achieve this. Central to
this strategy is the belief that it would take too
long to develop an electronics industry in Scotland
through indigenous growth and consequently it has
focused on attracting inward investment into
Scotland, which it pursues through the 'Locate in
Scotland' scheme.

Within their own terms, the policies pursued by
the Irish and Scots have undoubtedly proved
successful. Whether they achieve any wider impact on
the economies concerned, for example, in terms of
technology transfer, however, remains open to
question. Because of the international division of
labour which has emerged in the industry which has
enabled companies to separate out and optimally
locate the different stages of the production
process, many of the inward investments have been
production-only branch plants. In Ireland a number of
these have been little more than low value-added
assembly plants relying almost completely on
imported components (which ironically can exacerbate
balance of payments problems). The American computer
company, Amdahl, for example, established in Ireland
in 1978 a plant which produces 40% of the Company's
world-wide main-frame computer output, serving
markets in Europe, Canada, the Middle East and the
Pacific Basin. It achieves this production with just
300 employees (soon to increase to 500); there is no
R and D activity at all in the plant, and marketing
of the Factory's output is carried out from an office
in London (Foley, 1985).

Even without the doubts that these characteristics
raise concerning the limited impact of such
inward investment - in terms of technology transfer,
spin-off into SMEs and so on - the changes outlined
in the previous sections are likely to make inward
investment-led strategies for less favoured regions
increasingly difficult to implement successfully.

ii **The 'National Champion' Strategy:** Attempting to
build up one or two companies in order to achieve the
scale economies needed to compete with the TNCs would
be unlikely to succeed, given the developments
towards internationalisation of markets and
increasing scale noted earlier. A more realistic

variant of the strategy would be to focus on the home market, and to develop companies which are essentially 'systems integrators' - i.e. to create a strong national market presence based on a wide product range through a combination of their own design and manufacturing expertise and tied OEM (original equipment manufacturer) agreements coupled with their own label sales. To the consumer, the diversity of product sourcing would remain transparent. Although such a strategy might build up the image of a vigorous indigenous IT industry, there is of course the real possibility that the company becomes little more than a market responsive selling vehicle, only retaining the technical expertise needed to cope with the idiosyncratic nature of national standards and protocols.

iii **The 'Niche Market' Strategy:** The third strategy attempts to exploit the new opportunities associated with internationalisation of IT products. As the major producers concentrate upon achieving global product reach in their major markets, they create niche market opportunities for smaller companies. These niches themselves may be international in scale, and exist even in the semiconductor market. One such niche market noted earlier was the design of novel but dedicated circuits for specific applications using standard merchant chips; there are gaps between the standard chips and custom chip markets, arising from a demand for custom semi-conductors but only at low volumes, and this has created a niche market for the design of novel circuits (which cost only £2,000/£3,000 to design, compared with in excess of £40,000 for a custom chip; CURDS, 1985). In the software sector, niche markets are considerably more in evidence, and many new firms are formed to exploit application specific niche markets, often existing on a national scale.

A range of strategies can be devised to identify and exploit niche markets as a viable means of encouraging indigenous IT production. One such strategy employed in Ireland is to identify national market niches for the supply of components to the foreign-owned IT sector. A new division within the IDA has been created to tackle in a systematic way the constraints which have hindered the entry of domestic firms into the sub-supply market. The approach of this National Linkage Programme will be to work closely with selected local suppliers to help them develop the necessary expertise to meet the purchasing requirements of the larger companies, which usually demand more rigorous technical specifications than domestic firms can attain without substantial help (Foley, 1985). A further strategy being implemented in Ireland is the National Software Centre, which aims to help indigenous software companies to identify market niches and provide the technical, training, financial and other assistance to help them to exploit the opportunities identified.

Overall, we would suggest that a niche market strategy based on the indigenous sector, particularly in software, would seem to be the most appropriate strategy for regions with limited involvement in IT production.

Whichever strategy (if any) is adopted, however, it is important for it to be pursued in a vigorous and co-ordinated manner. In this respect, the contrasts between the situation prevailing at national level in most European countries with those of Japan and the US are instructive. Both of the latter countries have pursued strategies of industrial targeting with respect to the IT sector, led by the Department of Defence in the US case and by the Ministry of International Trade and Industry in Japan. In the UK national context (quite apart from any regional policy considerations) the present government's 'non-intervention strategy', fuelled by an ideological devotion to free trade and to the de-regulation of markets - will from a position of competitive weakness almost invariably hasten the decline of the UK's indigenous IT production industries. Perhaps we will achieve the distinction of being the first nation to have had the sun set on its sunrise industries?

References
Buswell, R.J. (1983) Research and development and regional development: A Review. In Technological Change and Regional Development, A Gillespie (ed.) London: Pion
Centre for Urban and Regional Development Studies (CURDS) (1985) The location of new information technology production in the UK. CURDS, University of Newcastle upon Tyne, UK
Dosi, G. (1981) Technical change and survival: Europe's semiconductor industry. Sussex European Research Centre, Paper 9, University of Sussex
Estell, R.C. (1985) Stock control in manufacturing. The just-in-time system and its locational

implications. Area, 17, 2

European Commission (1985) The Mid Term Review of ESPRIT. Report submitted by the ESPRIT Review Board, Brussels

Foley, A. (1985) Regional location of information technology and production in the Community. Case Studies of East and West Regions in Ireland. CURDS, University of Newcastle upon Tyne

Freeman, C. (1974) The Economics of Industrial Innovation, Harmondsworth, Penguin

Frobel, F. Heinrichs, J. and Kreye, O. (1980) The New International Division of Labour. Cambridge, Cambridge University Press

Gillespie, A.E., Goddard, J.B., Robinson, J.F., Smith, I.J. and Thwaites, A.T. (1984) The effects of new information technology on the less favoured regions of the Community. Commission of the European Communities, Regional Policy Series No. 23, Brussels

Goddard, J.B., Thwaites, A.T. and Gibbs, D. (1986) The regional dimension to technological change in Great Britain, in Technological Change, industrial restructuring and regional development, A. Amin and J.B. Goddard (eds), London: Allen and Unwin

Goldman, A. (1982) Short product life cycles: implications for the marketing activities of small high technology companies. R&D Management, 12, 81-9

Howells, J. (1984) The location of research and development: some observations and evidence from Britain, Regional Studies, 18, 13-29

McKinsey & Co. (1983) A Call for Action: The European Information Technology Industry. Report to the Commission of the European Communities, Brussels

Malecki, E.J. (1980) Corporate organisation of R&D and the location of technological activities. Regional Studies, 14, 219-34

Malecki, E.J. (1986) Technological imperatives and modern corporate strategy, in Production, Work, Territory: The geographical anatomy of industrial capitalism, A.J. Scott and M. Storper (eds), Boston: Allen and Unwin

Massey, J. (1984) Spatial Divisions of Labour: Social structures and the geography of production. London, Macmillan

Morgan, K. and Sayer, A. (1983) The international electronics industry and regional development in Britain. University of Sussex, Urban and Regional Studies Working Paper 34

Norton, R.D. and Rees, J. (1979) The product cycle and the spatial decentralisation of American manufacturing, Regional Studies, 13, 141-51

Oakey, R.P. (1984) High technology small firms:

innovation and regional development in Britain and
the United States, London: Frances Pinter
Oakey, R.P., Thwaites, A.T. and Nash, P.A. (1980) The
regional distribution of innovative manufacturing
establishments in Britain. Regional Studies, 14,
235-53
Oakey, R.P., Thwaites, A.T. and Nash, P.A. (1982)
Technological change and regional development: Some
evidence on regional variations in product and
process innovation. Environment and Planning A, 14,
1073-86
Palloix, C. (1975) The internationalisation of
capital and the circuit of social capital. In,
International Firms and Modern Imperialism, H.
Radice (ed.), Harmondsworth, Penguin
Porter, M.E. (1980) Competitive strategy, New York:
Free Press
Porter, M.E. (1983) Cases in competitive strategy.
New York, Free Press.
Rada, J. (1982) Structure and behaviour of the
semiconductor industry. New York: United Nations
Centre on Transnational Corporations
Savey, S. (1983) Organisation of production and the
new spatial division of labour in France. In, Spatial
Analysis, industry and the industrial environment.
Volume 3, Regional Economics and Industrial Systems,
F.E.I. Hamilton and G.J.R. Linge (eds). New York,
Wiley
Sayer, A. (1985) Industry and space: a sympathetic
critique of radical research. Society and Space, 3,
1, 3-29
Sayer, A. (1986) Industrial location on a world
scale: the case of the semiconductor industry. In,
Production, Work, Territory: The geographical
anatomy of industrial capitalism, A.J. Scott and M.
Storper (eds), Boston: Allen and Unwin
Storper, M. and Walker, R. (1983) The theory of
labour and the theory of location, International
Journal of Urban and Regional Research, 7, 1-41
Thwaites, A.T., Oakey, R.P. and Nash, P.A. (1981)
Industrial innovation and regional development.
Final Report to the Development of the Environment.
Centre for Urban and Regional Development Studies,
Newcastle University, UK.
van Tulder, R. and Junne, G. (1984) European
Multinationals in the Telecommunications Industry,
University of Amsterdam
UNIDO (1981) Restructuring World Industry in a Period
of Crisis - The Role of Innovation - An Analysis of
Recent Developments in the Semiconductor Industry,
Vienna: UNIDO.

Chapter Six

HIGH TECHNOLOGY INDUSTRY IN CANADA: LOCATIONAL AND
POLICY ISSUES OF THE TECHNOLOGY GAP

John N.H. Britton

Introduction
Although Canada's comparative advantage since the
foundation of the country has resided in its natural
resources, Canadian economic development policy over
the past century has been primarily devoted to the
growth of industrial jobs. Partly as a result of this
policy, its North American location, and its large
and continuous growth in market size, Canada leads in
industrial development among the group of resource-
based semi-industrial countries that have achieved
high per capita incomes. The industrialisation of
Canada, particularly after the Second World War, came
easily with direct investment spilling over from the
U.S. and with a high rate of international
immigration driving Canadian demand for industrial
production. Over recent decades, however, this form
of 'industrialization by invitation' (Naylor, 1975)
has proved to be inherently flawed as Canada has had
to recognise and try to cope with a serious and
growing technological gap between new industrial
performance and that of her competitors from the
developed world (Britton and Gilmour, 1978; Britton
1980, Science Council of Canada, 1984, 8).
      Canada now faces significant problems of
industrial structure and performance many of which
have their origin in the shallow form of industrial
development that has followed from a high level of
foreign direct investment in branch plant
operations:

-      The vertical integration of Canada's resource
       industries is weak, neither the supply of
       capital equipment nor downstream processing
       being highly developed.
-      The high level of U.S. ownership of branch
       plants in secondary manufacturing has proved to

be an ineffectual basis for technological development and for achieving manufactured exports commensurate with industrial production or employment.

- The high concentration of Canadian trade with the U.S. occurs despite the non-tariff barriers erected against imports into the U.S. but Canada is vulnerable against the rise in U.S. protectionist sentiments which now are being expressed not only at the state level but also in Congress.

- Canada has failed to develop coherent, consistent, and effective policies for the technological development of the industrial sector. In particular, Canada has found it politically impossible, in an era of economic growth (though lagging technological performance), to confront the priority of economic development over regional stabilisation and likewise it has failed to take adequate stock of its developmental potentials, particularly its locationally definable strengths, and to use them as the focus of policies that enhance the likelihood of industrial innovation and technological development.

With these problems as a context, three purposes direct the structure of this chapter, first, the background to Canada's problems in technological development is established in more detail. In particular, the small scale of Canadian high technology production is set within the framework of Canada's international trading pattern and the recurring issue of a trade agreement with the U.S. Second, the locational structure and changes in Canadian high-technology industry are identified and the dominance of the Toronto region rather than Silicon Valley North (Ottawa region) is established. Third, the incompatibility of these trends with federal and Ontario government programmes is assessed.

The Background to Problems of Technological Development

Despite the technological problems facing the Canadian economy, it is appropriate to note that the recent Canadian industrial record contains notable achievements by high-technology enterprises. The 'Challenger' executive aircraft, the DASH-7 and -8 STOL aircraft, and the CANDU nuclear power stations

are technically succesful products, but they have not
captured substantial international markets. On a
less pessimistic note, the remote manipulator arm
used on the U.S. space shuttle, was designed and
manufactured by a Canadian company, Spar Aerospace,
and this technology may eventually produce market
spin-offs. Government support of major proportions
has been provided to allow crown corporations and
private sector businesses to develop these product
systems based upon new technology; unfortunately
expected international market demand has either
contracted (due to recession and/or competition) or
the value of the investment has yet to be proven.
Canada's successes in digital switching, telecommunications systems, communications satellites, and
urban transit systems, however, indicate the
country's ability to bring high-technology products
into international trade. In these cases, too,
government assistance supported R&D in existing
firms and in the building of core companies so that
complex developmental and marketing tasks could be
undertaken. In other instances, businesses have
established themselves in international markets
without, or with modest, government assistance – one
example of many being the international computer
communications network of I.P. Sharp.

Despite these visible examples of high-
technology output by Canadian industry a more
accurate assessment would point to the serious
technological gap that exists between Canada's
industrial performance and that of its competitors
from the developed world (Britton and Gilmour, 1978;
Britton, 1980; Science Council of Canada, 1984, 8).

Canadian R & D. Seven industries in Canada are
usually included in the 'high technology' category by
virtue of their research intensity (R&D expendit-
tures/sales, Table 6.1). Some, like aircraft and
parts and communications equipment, undertake levels
of R&D that are well above the average but they are
as susceptible to cyclical factors as other sectors
and over the decade to 1982, R&D expenditures in
these industries generally sagged before recovering
and reaching new, higher levels (Table 6.1).
Engineering and Scientific services are an exception
to this pattern since they enjoyed a continuous
increase in research intensity throughout the period
because of expanding domestic markets for
technological service inputs (manufacturing and
service industries) and domestic and international

145

markets for engineering consultant services.

Table 6.1: Canada: Research Intensity of Selected
Industries Current Intra-Mural R&D Expenditures/
Sales

	1982	1982: 1973
Manufacturing		
Business Machines	2.2	0.9
Other Machinery	2.2	1.8
Aircraft Parts	14.5	0.9
Communications Equipment	10.6	1.7
Drugs & Medicine	5.2	1.2
Sci & Prof. Equipment	2.2	1.4
Total	1.3	1.4
Services		
Engineering & Scientific Services	10.9	2.3
Total Services	0.9	1.5
All Industries	1.2	1.2

Source: Statistics Canada, <u>Industrial Research and
Development Statistics 1983</u> Cat. No. 88-202.

The total level of R&D activity in Canadian
industry is dominated by the communications industry
and this is true even when other major, but less
intensive, R&D performers are included in the
comparison (Table 6.2). This industry is
distinctive, too, in the high proportion of its R&D
conducted by Canadian companies. Other industries
have increased their Canadian share of R&D - aircraft
and parts, for example, shifting from 7 to 50 per
cent Canadian control through government purchase of
foreign subsidiaries. The other significant shift
has been the decline in Canadian control of R&D in
the business machines industry from 40.5 to 3.6 per
cent which is explained by foreign take-overs. (Table
6.2). Canada has also attained a substantial scale of
R&D in electrical power generation and in the
petroleum industry; the former is Canadian (state)
owned while the latter is almost exclusively the
preserve of the giant multinationals. There is,
however, potential for future growth in the research
intensive industries if the current number of R&D
performing enterprises is any guide; there are at
least 500 domestically owned firms undertaking R&D
and another 200 foreign controlled R&D performers
(Table 6.2). Other machinery, scientific and

Table 6.2: Total Intra-Mural R&D Expenditures: Canada

| | 1984 | | 1982 | | | |
| | $'000,000 | Per cent | Number of Performers of R&D | | Expenditures Canadian Controlled firms per cent | |
			Canadian	Foreign	1982	1982:1973
Research Intensive						
Business Machines	104	3.9	9	6	3.6	0.1
Other Machinery	78	2.9	93	33	58.9	1.4
Aircraft & Parts	269	10.1	4	7	50.2	7.2
Communications	749	28.0	76	30	76.0	1.1
Drugs & Medicine	53	2.0	14	24	25.8	1.2
Sci. & Prof. Equip.	26	1.0	30	13	44.1	0.9
Engineering & Scient. Services	82	3.1	172	6	90.2	1.5
Research Intensive	1361	51.0	398	119	-	-
Other Major R&D Performers						
Gas & Oil Wells	132	4.9	11	6	56.3	2.1
Other Electrical Prod.	106	4.0	41	28	41.7	1.9
Petroleum Products	189	7.1	5	7	0.7	7.0
Other Chemical Prod.	125	4.7	49	56	34.5	1.3
Electrical Power	157	5.9	5	0	100.0	1.0
	709	26.6	111	97	-	-
Total	2673	100.0				

Source: See Table 6.1

professional equipment, and engineering and scientific services are populated by active clusters of Canadian enterprises from which threshold firms (Steed, 1982) may emerge.

High-Technology Production

The research intensive industries as a group (identified in the tables above) contribute only 13 per cent of Canadian manufacturing value added and, although the importance of this 'sector' has increased in recent years, the shift from 10 per cent value added is no less modest than that measured by employment or number of establishments (Table 6.3). In terms of vitality, when compared with the performance of all manufacturing, high-technology industry generated establishments and employment at nearly four times the rate for all manufacturing at the national level, and it is the small scale of the sector that has been the major constraint on the impact of its growth on the structure of Canadian manufacturing. Nevertheless, the implication of this superior performance is that high-technology industries have a greater chance to produce new firms that are likely to grow.

SME vs. Large Establishments in Canadian Industrial Growth

Regional analyses of technology-based economic growth have shown that institutional factors, the dynamics of a small-firm population in specific activities, and the interdependence of small and large firms are the key components of growth 'phenomena' in both North America and the United Kingdom (Rothwell, 1984; Segal, 1983; Oakey, 1984; Saxenian, 1984; Steed and de Genova, 1983). In particular, these analyses make it quite clear that the most obvious evidence of new entrepreneurial interest and vigour in a particular industry or region involves substantial increases in the population of small firms. Thus, to evaluate the performance of technology-based industry in Canada, two sets of information are required.

1. Aggregate measures, over time, of the performance of technology-based industries in which large establishments/firms will potentially have a dominating influence.
2. Data that allow changes in the number of small enterprises and their production and employment

Table 6.3: Canadian High Technology Manufacturing 1975-1981

	No. Establishments		Value Added $'000,000		Employment '000		Per cent change		
	1975	1981	1975	1981	1975	1981	Estabs.	VA	Emp
Bus. Equip.	39	71	245	672	10	16	82	174	68
Other Machinery	1059	1549	1802	4018	83	92	46	123	12
Aircraft Parts	101	153	438	1716	22	39	51	291	75
Communic.	262	427	930	1935	42	46	63	108	10
Pharm.	134	134	457	970	15	16	0	112	9
Sci. & Prof.	314	537	368	820	18	23	50	119	28
Total	1909	2871	4240	10131	190	232	71	138	22
All. Manuf.	30100	35780	38716	78261	1743	1854	19	102	6
% High Technology	6.3	9.7	10.0	12.9	11.0	12.5			

Source: Statistics Canada Manufacturing Industries of Canada; Sub-provincial Cat. No. 31-209

impact to be established and compared with the performance of medium- and large-sized enterprises.

The data that are available to assess the extent to which SMEs (small- and medium-sized enterprises) have been responsible for growth are highly limited. Given the dangers of using incomplete information, as outlined by Storey (1983) for the United Kingdom for example, conclusions for Canada can be drawn only with considerable caution, but the greater survival of jobs in the SME sector is of importance for Canada's high-technology development, given the level of foreign ownership in this sector. The general vitality of small enterprises in creating jobs has meant increased public attention, a parallel institutional response, and more assistance for young and new high-technology enterprises which in the past have been inadequately assisted by government initiatives and which have been regarded as difficult clients by Canada's banks.

Although information for individual manufacturing industries is not available, a special tabulation of data from the Census of Manufacturers for 1974-1979 produced useful aggregate evidence (Layne, 1980), as follows:

-    68 per cent of the establishments existing in 1974 survived the next 5 years and 1979 employment associated with these establishments, was 91 per cent of that in 1974;
-    for the smallest size group of enterprises (less than 20 employed in 1974) only 59 per cent survived but the total level of 1974 employment was retained;
-    establishments with more than 100 employees in 1974 had a higher survival rate (90 per cent) but a lower (10 per cent less) rate of employment retention;
-    although the number of establishments in the smallest size group retained its 62 per cent share over the period of manufacturing employment increased by only one percentage point to 7 per cent.

The significance of Layne's data is that they show in a period of capital intensification, rationalisation and job shedding by larger enterprises, that new small businesses continued to provide opportunities for modest growth in employment and therefore, a political base for

improved support systems. In fact, the situation can
be viewed even more positively when the longitudinal
data are examined in more detail: the growth
performance of the surviving small firms shows an
increase of 141 per cent (current $) in value added
compared with 73 per cent for manufacturing units
employing 100 or more. The implication, therefore, is
that small firms have been superior performers.

In order to bring the account up to date use is
made of two special studies that are based on non-
census business records. The Canadian Federation of
Independent Business (CFIB) tracked the employment
of its 1975 members who survived in 1982 and found
that the smallest firms show impressive growth
relative to their base size (Table 6.4) and,
furthermore, firms with less than 20 employees
generated the bulk of jobs.

Data drawn from Dun and Bradstreet credit files
for 1974-1982 include not only net employment change
for the base population of firms but also the impact
of subsequent growth. These data allow the CFIB
results to be amplified and supplemented, especially
for large enterprises which are not CFIB members.

Table 6.4: Canada: Job Creation 1975-1982 SME Members
of CFIB

	Per cent increase in employment	Percentage of net employment growth		Percentage of 1975 employment
0-4	75.7		36.8	6.1
5-9	28.2		25.9	11.4
10-19	19.1	(10-14)	14.2	9.0
		(15-19)	9.5	6.4
20-49	9.3		17.4	23.5
50-99	1.9		2.3	15.5
100+	-2.7		-6.1	28.1
Total SME	12.5		100.0	100.0

Source: Canadian Federation of Independent Business,
A Study of job creation 1975 to 1982 and forecasts to
1990 (1983).

According to the Dun and Bradstreet data 531,000
jobs were created in the manufacturing and service
sectors 1974-1982, dominantly by very small and very
large enterprises, medium-sized businesses being
poor contributors (Table 6.5).

At the level of 2-digit industries four patterns

Table 6.5: Canada: Employment Growth 1974–1982

	0–19	20–49	Employment Size of Enterprise 50–499	500+	Total
Manufacturing	82,929	18,640	-75,511	86,034	112,101
Service	191,973	55,238	67,413	105,080	419,704
Total	274,902	73,878	-8,098	191,114	531,805

Source: Dun and Bradstreet credit files processed for Department of Industry and Trade, Province of Ontario.

of change can be identified. For industries based on <u>traditional technology</u> employment in large enterprises declined substantially because of severe competition from imports supplying mass market goods. Employment in small enterprises was able to expand, though there are differences between leather, rubber, textiles, knitting, clothing and furniture in the degree to which they compensated for the employment losses of large enterprises (Table 6.6).

The capital intensive <u>resource processing</u> industries experienced substantially greater employment growth (as high as tenfold in the largest sized category) than the smallest, and medium-sized enterprises lost employment. Paper, primary metals, non metallic minerals, wood, tobacco and petroleum are included, although the latter actually lost employment among large producers. The remaining industries include <u>technology intensive</u> and <u>medium technology</u> activities; in the case of transportation equipment, for example, very large scale auto assembly plants (medium research intensity) dominate the influence of aircraft and parts which is the research intensive component of the industry group. In several industries large plants do not dominate employment and a small plant sector has always been important; this applies to printing, metal fabricating, and miscellaneous in which employment increases among small enterprises have been three times those of large establishments. By contrast in <u>technology-intensive</u> activities, in which large enterprises have dominated employment (frequently foreign-owned enterprises), employment increases among the very small firms <u>matched</u> those of large enterprises and so there are important clues, even in aggregate data, to organisational changes behind the pattern of employment increase.

Although the impression that is gained from the data presented so far is that Canada is experiencing high-technology growth, the fact remains that Canada has not gained technologically relative to its international competitors and evidence for this view is to be found in Canadian international trade data.

International Trade. Over the past decade, Canada's international trading pattern has shown a positive merchandise trade balance and, in contrast with industrial economies, has maintained a negative balance on services, resulting in an overall deficit. The merchandise trade surplus is produced as a result

Table 6.6: Canada: Manufacturing-net Employment Change by Enterprise Size, 1974–1982

Industry	0–19	20–49	50–99	100–199	200–499	500+	Total
Leather	987	674	-395	-1926	-1874	-2356	-4890
Rubber	3488	2063	1723	-1499	360	-4054	2081
Textiles	2709	1103	-1743	-1176	-1728	-2513	-3348
Knitting	633	278	-502	-1581	-2173	-779	-4124
Clothing	4571	2621	-2108	-5906	-6374	-3162	-10358
Furniture	4635	623	-1891	-3551	-2817	-1096	-4097
	17023	7362	-4916	-15639	-14606	-13960	-24736
Food	6242	684	-509	-2133	-1995	3576	5865
Paper	1133	856	22	-433	-543	10718	11753
Primary Metals	1636	535	-13	957	750	15612	19477
Non-metallic Minerals	2723	458	-881	305	-864	679	2420
Wood	7446	-408	-1941	-545	-3948	10621	11225
Tobacco	94	14	43	-100	-330	1753	1474
Petroleum	334	64	-87	-339	-309	-1089	-1426
	19608	2203	-3366	-2288	-7239	41870	50788

Size — Canada Totals by Manufacturing Group

Table 5.6: continued

Industry	0-19	20-49	Size Canada Totals by Manufacturing Group 50-99	100-199	200-499	500+	Total
Printing	9441	2311	-577	-433	-721	5507	15528
Metal Fabricating	11319	2629	-2079	-1904	-4883	2784	7866
Misc.	6034	-257	-728	-1380	-3522	586	733
	26794	4683	-3384	-3717	-9126	8877	24127
Machinery	7224	1938	270	-1880	77	4716	12345
Electrical	4297	2143	-467	-1937	-4531	3480	2985
Chemical	1781	280	-1089	-103	2184	5562	8615
	13302	4361	-1286	-3920	-2270	13758	23945
Transport	6202	40	-1049	-703	-2002	35489	37977
Totals	82929	18649	-14001	-26267	-35243	86034	112101

Source: Unpublished analysis, by Dr. David Birch for Department of Regional Industrial Expansion.

of surpluses for agricultural products, natural resources, raw materials, and energy products and a deficit for manufactured commodities. The manufactured trade deficit has fluctuated in the range of -2 to -4 per cent G.N.P. with the exception of 1982 when imports were severely curtailed during the recession (Table 6.7). It is important to note, however, that both manufactured imports and exports have been increasing relative to G.N.P (1982 again being an exception; Table 6.8) showing that Canadian manufacturers have had some success in international marketing despite the endemic deficit on manufactured trade balance.

Table 6.7: Canadian Merchandise Trade Balance/GNP Selected Years Per Cent

	Agri-cultural Products	Natural Resources and Raw Materials	Energy Products	Manufactured Products	Total
1971	0.32	2.37	0.40	-1.17	2.03
1972	0.31	2.50	0.59	-2.29	1.10
1974	0.54	2.53	1.15	-4.10	0.11
1975	0.41	1.87	0.73	-4.17	-1.16
1981	0.35	2.32	0.65	-2.15	1.16
1982	0.63	1.72	1.58	0.24	4.17

Source: Statistics Canada, 1983, Canadian Science Indicators, Cat. No. 88-201.

Table 6.8: Canada: Manufactures Trade, Selected Years

	Exports $'000,000	Imports $'000,000	Exports Per Cent G.N.P.	Imports Per Cent G.N.P.
1971	12,154	13,257	12.87	14.04
1972	13,533	15,946	12.86	15.15
1974	19,489	25,545	13.21	17.32
1975	20,583	27,486	12.45	16.62
1981	54,361	61,657	16.03	18.18
1982	54,957	54,101	15.41	15.17

Source: See Table 6.7.

High-technology commodities have major responsibility for holding manufactured trade in deficit although most manufactured goods do not achieve a

Table 6.9: Canada's High Technology Trade, Selected Years

	Imports $000,000	Exports	Trade Balance	Imports	Per Cent G.N.P. Exports	Trade Balance
1971	3,505	1,572	-1,933	3.71	1.66	-2.05
1975	7,319	2,909	-4,410	4.43	1.76	-2.67
1981	19,079	9,424	-9,655	5.63	2.78	-2.85

Source: See Table 6.7

surplus, the exceptions being resource products and motor vehicles (operating under the U.S.-Canada Automotive Trade Agreement). Furthermore, the trade deficit in high-technology goods has increased (relative to G.N.P.) as imports have expanded faster than Canadian high-technology exports (Table 6.9): 'the greater the amount of technology required, the greater our dependence on goods produced abroad' (Statistics Canada, 1983, 88-201, 95). Import penetration and export orientation have both been increasing within the high-technology sector (Table 6.10) but it should be noted that these technology-based industries are broadly defined and the levels of import penetration experienced by Canada are an important index of Canada's development problems.

Table 6.10: Canada: High Technology Industries: Imports and Exports 1981

	Imports		Exports	
	Penet-ration 1981	Per Cent 1981: 1966	Orienta-tion 1981	Per Cent 1981: 1966
Office Equipment	95	1.51	87	2.72
Office Machinery	73	1.14	50	1.52
Aircraft & Parts	66	1.65	62	1.29
Communications Equipment	59	1.79	49	3.06
Drugs & Medicines	15	1.00	7	1.00
Scientific & Prof. Equipment	81	0.96	54	0.73
	—	—	—	—
All Manufacturing	32	1.52	30	1.58

Source: See Table 6.7

Structural Problems of Canadian Industry. The explanations for Canada's weak development of high-technology industry are complex and continue to produce substantial disputes (Britton and Gilmour, 1978; Economic Council of Canada, 1983). Perhaps the most telling indication of the seriousness of the problem, however, is provided by Canada's failure to produce an industrial structure that gets full value (in terms of backward linkages) from the technological demands of its resource sector. In large measure, this has been a result of the high level of foreign ownership that has existed in the resource industries and which set in motion foreign sourcing of capital equipment in a pattern that has continued even though Canadian ownership in the resource sector is now greater.

Within secondary manufacturing Canada's problems involve both foreign ownership and an industrial structure which is itself an effect of foreign control as well as domestic factors:

1. In Canadian high-technology sectors large firms tend to be foreign owned, and neither as export oriented nor as R&D intensive as Canadian firms.
2. Foreign controlled firms dominate the output and sales of all high-technology industries, and demonstrate a much higher import propensity (sub-assemblies and components) than Canadian firms (Britton, 1985).
3. The small number of large Canadian high-technology firms, resulting from market fragmentation, means limited stimulation of smaller firms and fewer opportunities for new firms to be spun-off from well-established Canadian companies.
4. In Canada, the expansion of the information economy has been regarded as an economic change affecting mainly the service industries. In reality many information related jobs are found in manufacturing (in high-technology industries in particular) and thus Canada has no escape from the need to develop a stronger technological core to its manufacturing sector.
5. Despite large increases in employment, the quality of growth in the service sector is vulnerable. First, the producer services need not only demand from the service sector itself but also require a healthy manufacturing market in order to survive and expand. Second, the import of "invisibles" into Canada is very large and it is to be expected that Canada's payments to U.S. computer service firms for data processing, data bases and other services will increase as the power of international firms in developing the network market place is realised.
6. As noted, the non-merchandise trading account is in deficit. Partly this is attributable to interest payments and travel deficits but it also reflects the net outflow of dividend payments to foreign corporate headquarters, and embodies payments for technological and

business services mainly by foreign companies.
7.   It is necessary to remember that the world
trajectory of trade is one which accords
increasing importance to technology-based
products and systems but Canada's slow pace of
development has the effect of increasing its
technological gap. New technologies are rapidly
emerging that can transform existing industrial
processes and CAD/CAM, composite materials,
bonding technologies and biotechnology are the
obvious examples of ways in which new
international competence can be established,
essentially within the shell of existing
sectors. But here, too, opinion seems to suggest
that Canada is lagging even though some of these
new technologies can enhance the productivity
of Canada's staple industries.

Canadian Viewpoints on the Technology Gap. The period
prior to and during the Tokyo round of GATT (1978)
was one in which substantial attention was given to
the likely impact of free trade on Canada's trading
strengths and weaknesses (Economic Council of
Canada, 1975; Britton, 1978). In practice, the Tokyo
round produced a regime of declining tariff
protection but increasingly, since tariffs began
falling, it has been realised that U.S. non-tariff
barriers effectively constrain the sale of Canadian
products and systems in the U.S. if they are produced
without the benefit of a U.S. branch plant and/or R&D
facility!
       Recently the possibility of a trade agreement
with the U.S. has rekindled speculation about the
pattern of specialisation that could develop in
Canada's trade and industrial structure. Two schools
of thought have emerged. Basically the neo-classical
economists' position sees problems of trade being
resolved by the adoption of free trade, the process
being one in which the rationalisation of firms and
sectors (through a clarification of Canadian
comparative advantage driven by greater access to the
large U.S. market) would produce new specialisations, corporate successes, new levels of industrial
productivity and improved trade performance.
Recently, support has been given to this position by
the Royal Commission on the Economic Union and
Development Prospects for Canada. Alternatively it
is argued from the perspective of applied social
scientists that Canada's trade performance has been
heavily influenced by institutional and organisat-

ional factors which will continue to mediate the impact of 'freer' trade with the U.S. The foundations of Canada's commercial relations, according to this view, cannot be captured by an inherently limited model, and changes in these relations cannot be sensitively predicted given the implicit assumptions made by the modellers. Rather, Canadian technological development, whether there is some form of trade agreement with the U.S. or not, depends in each industry on the level of foreign control, the degree to which world product mandates have been developed by foreign-owned firms, on corporate policies for North American rationalisation, the degree to which Canadian-owned firms have been spending on R&D and marketing, and the extent to which there are technology-intensive Canadian firms that are poised for expansion of their sales in the U.S. Government policies on trade financing, support for new enterprise, international marketing and R&D, for example, are also important additional influences, and the recent replacement of a long-standing federal Liberal government by a Conservative administration (and the reverse in the most highly industrialised Province of Ontario) is likely to produce different arrangements for the assistance of technology-intensive firms.

Most of the influences outlined above are implicitly or explicitly geographic in their form – for example, many federal and provincial policies that currently operate to assist technological development have locational biases in their pattern of delivery (see below). Likewise, foreign-owned enterprises in high-technology activities are locationally concentrated in southern Ontario and the agglomeration factors most favourable to the generation of new technology-based enterprises are probably at their most developed in the Toronto region of southern Ontario. These locational aspects of Canadian development have also made southern Ontario the focus of the interprovincial tensions produced by Canada's type of industrialisation: it is both the industrial 'heartland' and the location of the primary impact of corporate rationalisation: its industrialisation has dominated the geography of Canadian manufacturing for a century but since 1978 it has been the location for much of the impact of declining tariffs.

The hegemony of southern Ontario is an important political force behind the support by other provinces for liberalised trade with the U.S., even though they are not equally persuaded about the benefits of free trade. Ontario's interests lie in minimising the risks involved in changing the rules of North American trade. The argument advanced by most provinces, other than Ontario, is that tariffs have produced a variety of inefficiencies which derive from the development of a domestic (east-west) system of exchange in contrast to a North American (north-south) establishment of provincial specialisations. It is precisely the domestic determination of specialisation - manufacturing at the centre - which has rankled, and based on the hindsight of this experience other provinces are supporting a new trading order which they hope will produce a more equitable distribution of high-technology activities which they expect to grow in employment. Nevertheless, this hope begs questions that hinge on the current location and recent changes in Canada's manufacturing industry (especially technology intensive activities) and its supportive ties with the service sector.

Location of Canadian Technological Activity

Because it includes business services a useful benchmark for subsequent locational analysis is derived from a 1977 Canada-wide survey of R&D establishments which recorded a 46 per cent response rate. Of the total respondents 33 and 18 per cent are in Toronto and Montreal CMAs respectively, while secondary nodes are Vancouver, Ottawa, Calgary and Edmonton (Table 6.11). The employment associated with the S&T units also places Toronto in primary place with 31 per cent followed by Montreal 26 per cent, Ottawa, Edmonton, Vancouver and Calgary. The shifts in rank order are attributable to the location of several large S&T units of manufacturing firms in Ottawa and of petroleum firms in Edmonton.

Apart from Toronto and Montreal only a few centres have developed a significant presence in Canadian technological activity. This pattern is highly localised because:

1. 85 per cent of the S&T employment is in the six largest centres accounting for 37 per cent of 1981 population, and
2. over one half of S&T employment is found in the two main centres with about one fourth of the total Canadian population.

The highest localisation (Table 6.11) is found in Ottawa where the attractive influence of federal

Table 6.11: Canada: Scientific and Technical Employment 1977

	S&T 1977 Empl. %	S&T 1977 Business Units %	1981 NSEM Empl. %	1981 Population %	S&T location quotient (pop. base)	Persons Engaged in R&D 1982 %	Intramural R&D expend. 1982 %
Toronto	31	33	18	12	2.6	25.2	24.3
Montreal	26	18	13	12	2.2	20.7	20.9
Ottawa	10	8	6	3	3.3	13.9*	13.7*
Vancouver	8	10	6	5	1.6	NA	NA
Edmonton	5	4	4	3	1.7	NA	NA
Calgary	5	6	7	2	2.5	NA	NA
Total	85	79	54	37			

*National Capital Region.

Source: MOSST (1978); Statistics Canada Cat. No. 88-202

government agencies has made a substantial impact on scientific and technological work in the private sector. Corroboration of these S&T patterns is provided by two additional types of information. Natural scientists, engineers and mathematicians who are employed in the public as well as the private sector, and who are, therefore, more widely distributed across the provincial capitals, have a similarly strong Toronto-Montreal bias. Data on R&D employment in the private sector, however, strongly favour Toronto and Montreal and in fact provide an indication of the location of the technological functions that are included in the S&T data together with business services.

The location of S&T employment reflects sectoral differences: through their size Toronto and Montreal exert a strong influence on the aggregate sectoral pattern and their specialisations are distinguished by high proportions of S&T employment in technology intensive manufacturing and Montreal has a higher than average proportion of employment in other secondary manufacturing and a comparatively low business service proportion (Table 6.12). Vancouver, Edmonton and Calgary, with small amounts of secondary manufacturing activity, are specialised in business services and resource related S&T employment, leaving Ottawa with high proportions of business service employment and utilities as a reflection of the development of government related activities including contract research. Toronto emerges as the centre with relatively high proportions of S&T employment in secondary manufacturing and in business services it is the largest concentration which is explained in terms of technical services to manufacturing, service sector firms and to the head office functions which cross sectoral boundaries. Toronto is the dominant head office centre for non financial corporations and a major centre for financial institutions.

Location of Technology-Based Manufacturing. Based on international experience, the location of production and the birth of new establishments in Canada's small high-technology sector should be highly localised, since production in Canada does not involve long production runs. In fact, the Silicon Valley North accolade has already been awarded by the Canadian media to the Ottawa region and it is important to understand that the area's technology-based growth is notable because of the visibility and rapidity

Table 6.12: Specialisation in Scientific and Technological Employment 1977 (per cent)

	Business	Utilities	Technology Intensive	Manufacturing Other Secondary	Resource and Primary	Total
Toronto	41	14	17	19	9	100
Montreal	32	13	20	26	9	100
Ottawa	61	14	5	17	3	100
Vancouver	51	1	14	16	22	100
Edmonton	56	11	6	6	22	100
Calgary	68	1	3	6	23	100
All Centres	44	10	15	20	11	100

Source: MOSST (1978).

with which it occurred in the 1970s in an urban
context in which there had been limited prior
industralisation. Steed and de Genova (1983)
identified 63 high-technology private sector
manufacturing firms in the Ottawa area in 1982 while
probably another 22 computer service firms could be
added to this group for a total employment of 15,000.
They acknowledge that some reports suggest as many as
300 firms but indicate that many are small branch
offices and service firms. The degree of
specialisation of Ottawa's high-technology manufacturing - electronics, telecommunications equipment,
avionics, scientific equipment and medical
instruments - is quite marked as is the pervasive
influence of the federal government in the
attractiveness of the location. The Departments of
National Defence, Supply and Services, and Industry,
Trade and Commerce and the federal National Research
Council represented powerful locational advantages
in the view of firms surveyed by Steed and de Genova.
But as important is the large scale of Control Data
Canada and Bell Northern Research and its
manufacturing affiliates because not only have they
made substantial impacts on the skills-structure of
the labour market but also they have spawned new
firms, the BNR complex itself generating 14
offshoots, partly as a result of the failure of one
of its affiliates.

Nevertheless, more substantial technology-based
developments did occur elsewhere, especially in
regions such as Toronto, illustrating that well
developed, even mature industrial regions, can
initiate technological development, Boston being the
archetype. Given the greater scale and industrial
depth of the Toronto region which gives rise to
market, labour, and technological advantages, it is
not surprising to find its recent growth in the
number of new technology-based firms outstripping
that of all other centres. Therefore, it is important
to establish and evaluate the patterns of location
and locational change as they occur across Canada for
technology-based economic activity. A variety of
data sources is employed, none of them perfect, but
collectively they are very useful in identifying
locational structures and trends at the provincial,
urban, and community scales.

Provincial Patterns of Location. The six high-
technology industries (defined at the 3-digit level)
provide the most complete indicator of the location

of technology-based manufacturing when considered at
the provincial level (Table 6.13). The level of
concentration of these activities in Ontario and
Quebec is very high - the lowest proportion being 75
per cent of Canadian value added for other machinery
which is the most widely located across the country.
Ontario alone accounts for more than half the value
added in each industry, except aircraft and parts and
Ontario produces 55 per cent of the total of high-
technology industries.

There were some changes in the location pattern
over the period 1975-82, however, and in addition to
shifts in the proportion of value added, there were
large increases in the number of establishments and
in employment. Using cut-offs of a shift of 2 per
cent points in value added, and net increases of 50
establishments and 2,000 employees to define growth,
most change is found concentrated in Ontario and
Quebec despite resource-based expansion in western
provinces (Table 6.14).

In the two industrial provinces, Quebec gained
dramatically in the production of aircraft and
business equipment and this occurred at the expense
of Ontario. Nevertheless, Ontario's strength in
communications increased and in scientific equipment
and other machinery both provinces experienced
substantial growth in number of establishments.
These shifts can be interpreted as gains made by
major corporate enterprises in Quebec while Ontario
emerged as the location favoured with substantial
employment increases and start-ups of new
enterprises. Location quotients for value added and
establishments confirm this interpretation for
communications and other machinery and in the
scientific instruments industry where small firms
predominate Ontario has quotients above 1.0, and
value added is highly localised.

Location within the Urban System

Given the Ontario and Quebec concentrations of high-
technology industry and the importance of Toronto and
Montreal in generating Canada's S&T workforce it is
no surprise that these two urban regions are the
focus for a very large proportion of technology-based
manufacturing. Analysis of patterns of location and
locational change are difficult because the
published statistical record of the Census of
Manufacturing becomes fragmented at the metropolitan
scale. Nevertheless, the six manufacturing
industries are maintained as the unit of analysis in

Table 6.13: Regional Distribution of High Technology Manufacturing

Region	Communications Equipment 1975	1981	Business Equipment 1975	1981	Scientific & Professional Equipment* 1975	1981	Aircraft and Parts 1975	1981	Drugs and Medicines 1975	1981	Other Machinery 1975	1981
Establishments												
Atlantic					36	56					23	33
Quebec	47	102	5	16	146	264	30	44	48	50	182	305
Ontario	160	238	31	46	324	459	36	49	76	70	575	791
Manitoba	6	14			35	31	5	10			45	58
Sask.											36	50
Alberta	9	17			83	101	9	16			76	157
B.C.	31	39			89	170	11	27			122	155
Employment '000												
Atlantic					1	1					1	2
Quebec	13	13	2	4	3	3	10	19	7	3	17	17
Ontario	25	28	7	10	15	18	9	16	8	8	49	54
Manitoba		1					2	3			5	6
Sask.											2	2
Alberta	1	1			1	1	1	1			4	7
B.C.	2	2			1	1					5	7

Table 6.13: continued

Region	Communications Equipment 1975	1981	Business Equipment 1975	1981	Scientific & Professional Equipment* 1975	1981	Aircraft and Parts 1975	1981	Drugs and Medicines 1975	1981	Other Machinery 1975	1981
Per Cent Canadian Value Added												
Altantic					1.6	1.6					0.6	0.6
Quebec	32.2	32.7	10.7	25.0	11.2	9.7	47.3	54.4	43.8	44.3	20.0	16.5
Ontario	58.1	58.2	89.0	72.8	79.1	79.2	40.8	37.3	54.6	54.4	62.3	58.3
Manitoba	0.9	2.1			1.1	0.7	7.4	5.5			5.3	7.1
Sask.											1.9	2.6
Alberta	2.6	2.2			2.8	3.7	2.4	1.0			4.1	8.8
B.C.	3.6	2.5			2.8	3.4		1.0			5.7	6.5

*Includes dental laboratories

Source: Statistics Canada Cat. No. 31-209.

Table 6.14: Canada: High Technology Growth 1975-1982

Province	Industry	Shift in Value added percentage 2%	Increase in net number establishments 50	Increase in employment 2,000
Ontario	Communications		78	3
	Business Equipment			3
	Scientific & Prof. Equipment		135	3
	Aircraft & Parts			7
	Other Machinery	4	216	5
Quebec	Communications		55	
	Business Equipment	14		2
	Scientific & Prof. Equipment		118	
	Aircraft & Parts	7		
	Other Machinery		123	9
Alberta	Other Machinery	5	81	3
B.C.	Scientific & Prof. Equipment		81	
	Other Machinery			2

Source: See Table 6.13

as complete a fashion as possible. Generally Census Metropolitan Area definitions are employed augmented with comparable statistical aggregations when necessary.

Three patterns of high-technology location can be identified.

1. The wide diffusion of <u>other machinery</u> through the Canadian urban system (Table 6.15), in accordance with the resource and industrial base of Canada, is expected given the provincial patterns discussed above.

    Although Toronto is the main centre (18 per cent value added) four other locations - Montreal, Edmonton, Hamilton and Vancouver - contribute at least 5 per cent each. Toronto and Montreal are notable because of their generation of new establishments.

2. The dominance of Toronto and Montreal in the location of all other high-technology industries: Toronto with nearly 36 per cent and Montreal with nearly 32 per cent. In absolute terms, therefore, other locations are of very limited individual importance and in fact only Ottawa with 19 per cent of business equipment value added (compared with Montreal's 20 per cent) is a competitive location. There have been small shifts in the importance of Toronto and Montreal over the 1975-1982 period - the share of aircraft production has contracted slightly in Toronto and expanded in Montreal though Toronto's employment has grown by 5,000.

- Toronto's proportion of business equipment production declined from 38 to 31 per cent but its employment increased.

- In communications, Toronto contributed 30 per cent of value added and generated net new establishments implying its potential for growth. The industry is widely distributed, however, 38 per cent of value added being located in Ontario centres other than Toronto and Ottawa (5 per cent). Montreal is the second location with 24 per cent of value added.

- In scientific and professional equipment nearly 55 per cent value added occurs in Toronto. Between 1975 and 1981 there was a surge in the net number of establishments in several locations especially in Montreal and Toronto.

    The relative as well as absolute strength of Toronto and Montreal is substantial and with few exceptions (Table 6.16) all high-technology

Table 6.15: Canada: Location of High Technology Industries 1975, 1981 Selected Locations*

Location	Establishments 1975	Establishments 1981	Employment '000 1975	Employment '000 1981	Value Added $'000,000 1975	Value Added $'000,000 1981	Per cent Canadian Total 1975	Per cent Canadian Total 1981
Other Machinery								
Toronto	264	358	16	16	383	712	21.3	17.7
Montreal	108	173	12	11	233	518	12.9	12.9
Edmonton	35	60	2	3	38	206	2.1	5.1
Hamilton	36	56	3	5	63	243	3.5	6.1
Vancouver	100	127	4	5	84	221	4.7	5.5
Business Equipment								
Toronto	19	23	4	4	93	207	37.9	30.8
Montreal (est)	NA	24	NA	2	NA	133	NA	19.8
Ottawa	6	10	1	1	28	127	11.2	18.9
Communications Equipment								
Toronto	90	140	12	13	278	570	31.8	29.5
Montreal	60	83	8	10	251	452	27.0	23.7
Ottawa	17	31	2	2	38	92	4.3	5.0

Table 6.15: continued

Location	Establishments 1975	1981	Employment '000 1975	1981	Value Added $'000,000 1975	1981	Per cent Canadian Total 1975	1981
Drugs and Medicine								
Toronto	54	43	6	6	183	374	39.9	38.6
Montreal	41	42	6	7	193	393	42.2	40.5
Scientific and Professional Equipment								
Toronto	167	257	9	12	240	496	58.2	54.9
Montreal	79	180	1	2	10	59	2.4	6.5
Ottawa	24	26	1	1	17	67	4.1	7.4
Aircraft and Parts								
Toronto	22	28	8	13	160	580	36.6	33.8
Montreal	NA	37	NA	19	NA	928	NA	54.1

* A minimum of 5 per cent 1981 value added

Source: Statistics Canada, Cat. No. 31-209.

173

sectors record value added location quotients greater than 1. These centres have strong, broad profiles of high-technology competence, Montreal's strength being more sharply focused.
3.   In the patterns of localisation of individual industries, Ottawa stands out as a centre that has achieved a remarkable profile of specialisation (Table 6.16) particularly in business equipment, scientific and professional equipment and communications equipment.

Table 6.16: Canada: Localisation of High Technology Industries

Industry	Value Added Location Quotients, 1981		
	Toronto	Montreal	Ottawa
Aircraft and Parts	1.7	3.7	–
Business Equipment	1.6	1.3	24.1
Scientific & Prof. Equipment	2.8	0.4	9.4
Communications Equipment	1.6	1.6	6.3
Drugs and Medicine	2.0	2.7	2.5
Other Machinery	0.9	0.9	–

Source: Statistics Canada, Cat. No. 31-209.

Nodes of New Technology Based Growth

The popular description of the Ottawa region as Canada's Silicon Valley is to a large degree misplaced because it overlooks Toronto's dominating position whether measured in terms of scale of production (value added) or enterprise formation. Nevertheless, the Toronto region is an agglomeration of smaller communities in which the City of Toronto and a surrounding cluster of boroughs and former boroughs (comprising Metropolitan Toronto) is the core. On the periphery of this region and among a variety of industrial nodes, old market towns, and dormitories, two communities - Mississauga in the west and Markham to the north-east have emerged as high-technology nodes. Attaching measures to this observation, though it is difficult because of the lower (2-digit) level of detail that is available from the Census of Manufacturers, is an appropriate task because Mississauga and Markham grew at about the same time as Ottawa.
    If comparisons are limited to the machinery and electrical products sectors, because of the

Table 6.17: Industrial Change in Toronto Communities

Location	Establishments		Employment		Value Added $'000,000		Per cent Canadian Total	
	1975	1981	1975	1981	1975	1981	1975	1981
Electrical Products								
Ottawa*	17	31	2162	2671	39697	95897	1.9	1.9
Mississauga	26	40	3232	3242	67293	162700	3.2	3.2
Markham	14	24	984	1447	25379	50349	1.2	1.0
Machinery								
Ottawa	6	15	1208	825	27534	127787	2.1	2.7
Mississauga	47	83	3389	4971	87190	233262	6.5	5.0
Markham	15	28	490	555	10452	32308	0.8	0.7

* Different areal definition from that used in Table 6.15 has excluded only a few very small establishments

Source: Statistics Canada. Cat. No. 31-209.

localisation of the technology intensive components that they contain, Mississauga is found to be a more significant location than Ottawa (Table 6.17). Over the period 1975-81 this difference was maintained in electrical products and only in relative terms did Mississauga's share of the machinery industry decline: in 1981 it had nearly twice Ottawa's value added and six times its employment in machinery. Over the period Mississauga established a strong increase in employment while Ottawa added just over 100. Perhaps as distinctively, Mississauga generated 'new' establishments and this characteristic is held in common with Markham which by 1981 also had more establishments than Ottawa, though the majority were substantially smaller. Whether this recent growth is primarily a result of the formation of new enterprises, the establishment of spin-off firms or of branch plants is a research question that has yet to be answered.

Some measure of changes that have occurred more recently is obtained from Scott's industrial directory (Table 6.18), which allows a tabulation of net changes in the number of establishments responsible for relevant product-lines. Not only is growth in activity localised within the Toronto region - Mississauga and Markham - but new product lines are focused within the computing and communications fields. These two locations appear to be continuing their more rapid agglomeration of enterprises than the Ottawa area. It is acknowledged that these data are crude because they ignore size differences among enterprises and between product lines but they are effective in suggesting the direction that more research should take in order to establish the comparative industrial depth of high-technology growth in these locations.

Other High-Technology Activities
Although most attention here has been devoted to the manufacturing sector, at several stages it has been made clear that 'business services' are an equal generator of jobs when data on employment is considered. Their importance is underlined by their high rate of growth 1971-1981 which at 130 per cent outstripped that of the more modest technology-intensive industries.

Business services are comprised largely of engineering and scientific services and computer services. The engineering related activities have developed a diffuse locational pattern as specialist

Table 6.18: Establishments in the Toronto Region: Selected Product Lines

		Metropolitan Toronto	Mississauga	Markham	Other Toronto Regions	Ottawa
Biological Products	1978	9	3		1	1
	1984	12	2	2	1	1
Inorganic Chemicals	1978	32	6	1	5	1
	1984	34	10	2	4	
Pharmaceutical	1978	47	10		4	1
	1984	44	18	4	5	
Aircraft Parts & Rel. Equip.	1978	29	8		2	3
	1984	21	15	1	5	1
Semi Conductors & Rel. Elec. Components	1978	14	3		3	2
	1984	16	7	3	2	8
Electronic Computer Equip.	1978	21	5	1	1	8
	1984	32	20	12	8	14
Automatic Merchan. Calculating/Accntg Machines	1978	9	1			1
	1984	6				1
Telephone & Telegraph Equip.	1978	11			3	5
	1984	14	1	3	1	7
Radio/TV Trans. Signalling/Detection Equip.	1978	48	8	6	3	13
	1984	75	30	11	12	24
Totals	1978	220	44	8	22	35
	1984	254	103	38	38	56

Source: Scott's Directories (1985) Ontario Manufacturers (16th ed.).

177

firms have emerged to serve the resource, construction, manufacturing and consultant markets and as these are often regionally defined the relative breadth of dispersion results (Table 6.19). By contrast, computer services are localised in Toronto and Ottawa and reflect the location of the computer software industry as well as computer service bureaux. The development of the software industry is, in many ways, substantially less impeded in its international sales by tariff and non-tariff barriers. If the 'computer industry' is considered in very broad terms, it is much more highly developed in telecommunications software, and in consulting than in computer hardware design and production. With no census providing business data on the software portion of the industry it has been difficult to establish its size, structure and location without painful recourse to industrial directories. At this time, however, data bases are becoming available for government use and locational research will soon be possible.

Table 6.19: S&T Employment in Business Services 1977

	Computer Services	Engineering and Scientific Services	All Business Services
Toronto	38.7	26.4	29.2
Montreal	8.8	20.3	18.8
Edmonton	1.6	7.0	6.0
Calgary	11.4	7.0	7.4
Vancouver	3.4	9.0	9.0
Ottawa	25.0	10.8	12.5
All Centres	88.9	80.5	83.9

Source: MOSST (1978).

Canadian Policies for High-Technology Industry

The recent edition of Assistance to Business in Canada (Federal Business Development Bank, 1984) contains nearly 200 pages of programme specifications, many being concerned with encouraging R&D, production expansion, and marketing by high-technology industry. From the standpoints of industrial firms and analysts alike the policy environment is exceedingly complex because of the lack of a coherent technological strategy. This issue has been the subject of substantial commentary in the past (Britton and Gilmour, 1978; Britton, 1985) but there has been little real response to the call to simplify and structure Canada's policies directed towards technological development. As noted earlier, the need for a technological strategy hsa been contested by some economists who not only believe in the efficacy of 'the market process' (especially if government intervention were reduced) but also see no place for national R&D or technological goals or for concern about the high level of foreign direct investment in Canada (Rugman, 1981; Daly, 1979; Safarian, 1979). The debates have not produced a substantial shift in federal policies - other than raising the national R&D/GNP target - probably because of a variety of regional and industrial interests that have pitted the resource-based economies of the eastern and western provinces against the industrial centre. The vulnerability of traditional, and some medium-technology, manufacturing industries has been important and the Tokyo round of GATT and the reduction of tariffs at a time of recession has, of course, intensified their problems. By contrast, the provinces, especially Ontario have recognised more clearly the need for a technological strategy as they try to deal with a new trading order.

Despite the lack of coherence in federal technological policies it is possible to enumerate a variety of reasons for the initiatives that exist:

1. Firms have difficulty appropriating all of the potential gains from their R&D expenditures and as a result may spend less than is socially desirable. The expected rate of private economic return can be raised by means of incentive subsidy payments.

2   R&D projects are inherently risky and firms may be unable to spread this risk over a large enough number of projects and thus grants, loans and/or equity participation is justified (Cohen, et al., 1984).

3. Risk factors associated with R&D are higher in the small economy/market faced by Canadian producers and furthermore, Canada has a low proportion of large domestic firms that are technology-intensive and small firms, presumably, are in greater need of assistance than successful international firms.

4. Canada's technology gap has been defined in terms of its international balance of payments

on current account but lag-times on technology adoption (modernisation) and low rates of innovation are also appropriate foci for policy measures.

5. Access to international markets may require concessional financing for purchasing firms/countries and Canada is justified in eliminating barriers to exports by matching the export financing available to competitors.

6. Risk assessment by traditional financial institutions may inhibit the banks making venture loans for small and untried technology-based firms even though it is in the public interest that working capital is available for these enterprises. By means of loan guarantees or by establishing a 'lender of last resort' the financial impediments to growth may be reduced. Similarly, venture capital (for equity) in technological enterprises can be made available by increasing the private rate of return on (risky) high-technology investments.

7. When the private sector has failed to provide a sufficient range and quality of producer services - product testing, technology advice, design, research into new technologies, for example - it has often been deemed desirable, from the standpoint of the national interest, to eliminate the supply-side failure through public investment. In Canada this is justified by the small size of the market for specialised industrial inputs and by the diversion of demand by foreign subsidiaries from potential Canadian suppliers to foreign parent corporations.

8. Government procurement, both for civilian use and defence, is commonly perceived as a means of providing a risk-free market for domestic enterprises and an opportunity for technology development.

Each of these arguments has elicited one or more responses at the federal and provincial levels and the forms of some of the policies that have been instituted are reviewed.

1. A number of programmes include support for high-technology industry, some are even targeted on the need to reduce risk and to offset problems of appropriability viz:

    i. under the Industrial and Regional Development Program (IRDP) of the

Department of Regional Industrial Expansion (DRIE) grant assistance is available
for innovation, restructuring modernisation and marketing. DRIE allocated nearly
$67 million (of a total disbursement of
$418 million in 1984-85) to innovation
projects.

ii. DRIE administers the Defence Industry
Productivity Program (DIPP) which spent
$169 million in 1983-84 on projects
directed to defence exports and technological capability - specifically defence
R&D. Currently DRIE is administering the
industrial spin-offs from a $600 million
contract for a low level air defence
system.

iii. The Industrial Research Assistance Program
(IRAP) of the National Research Council
(NRC) spent $48 million in 1983-84 on small
and medium sized manufacturing firms. The
cost of laboratory or consulting contracts
and the salaries of technical personnel to
solve specific technical problems are met
by this programme.

2. Loan programmes with a more general application
exist for small businesses undertaking
'modernisation' and exports but only the
Federal Business Development Bank has been
mandated to assist start-up firms. It also
arranges venture/equity financing. In Ontario,
Small Business Development Corporations direct
equity funds obtained through personal tax
concessions.

3. Technology transfer mechanisms have been
developed through NRC which operates 20 field
advisory service offices linked with the
Technology Information Service which solves
technological problems, improves production
systems and assists in the development of new
products or processes for small companies. In
Ontario, the provincial government has
established technology centres in microelectronics (Ottawa), robotics (Peterborough) CAD/CAM
(Cambridge), auto parts (St. Catherines), farm
equipment and food processing (Chatham) and
resource machinery (Sudbury). These are
diffusion centres for specific technologies:
all operate on a cost recovery basis as does the
Ontario Research Foundation which provides a
variety of testing, laboratory and research

services, especially for small- and medium-sized firms.
4. Federal research laboratories have been developed in a very broad range of fields over many decades and the NRC operates a programme designed to link industrial companies with these laboratories.
5. Canada's mainstay of technology support, despite the plethora of programmes, is a set of tax incentives which allow industrial firms to write off substantial portions of their R&D expenditures. Two procedures have been developed which in 1981 allowed $125 million in investment tax credits and $282 million of additional allowances. In both cases a minority of R&D performers claimed: presumably they did not achieve taxable status! The nature of the tax provisions for R&D which have operated to this time have been rated as second only to those of Singapore.
6. A variety of federal and provincial initiatives have been taken to create university-based incubators and high-technology firms. These are small ventures with limited impact at this time though considerable local energy has been spent which has probably made the accessibility of university consulting more feasible to firms. Attempts to link industry and the universities through research contracting have proceeded faster in the 1980s than previously.

While the catalogue of approaches to policy formulation might appear impressive the criticisms are equally definite. The notable feature of policy delivery in Canada is that it is overwhelmingly passive in its mode of approaching the problem of reducing the technology gap. Furthermore, the tax system is notoriously indirect and impossible to target because it relates directly to tax savings rather than to the proven ability to produce a commercial success as a result of R&D. The large proportion of non-claimant firms undertaking R&D is an indication of the effects of the recession and their ignorance of correct procedures (Steed, 1982) and the pattern whereby large firms receive most benefits reveals the poor design of Canada's major programme of assistance for high-technology firms. It has long been argued that poor granting procedures lead to 'windfall' gains by established R&D performers: the tax relief schemes are probably the worst offenders at giving benefit where none is

required. Over half the tax credits claimed accrue to
foreign owned firms and this is an ironical twist to
the outcome of policies concerned with national
development of technology-based industry. The
negative aspects of the way credits are taken up are
increased because of the narrow definition of R&D
that has been employed: experimental development
work has been excluded thus assisting the support of
research laboratories in multinationals which can
export their results while denying assistance to
Canadian firms, in the machinery industry for
example, that require support for product
development (Ondrack, 1980) and engineering follow-
through (Steed, 1982), common in the later stages of
the innovation process.

The lack of any active delivery of policy
instruments in Canada, however, is not restricted to
the tax system. It is characteristic, that even with
the loan/grant programmes that contacts are made
first by interested firms which tend to be the most
knowledgeable. Agencies then respond along technical
lines, as with the NRC and Ontario's technology
centres or, the granting bodies such as DRIE proceed
in a business appraisal mode. Generally NRC has won
positive evaluations from the Science Council
(Britton and Gilmour, 1978), Economic Council (1983)
and Wright Report (1984). IRDP and its predecessors,
however, have received a variety of negative
appraisals (Economic Council, 1983; Wright Report,
1984 and Britton and Gertler, 1985) mainly because of
the complex form of its mandate. The IRDP is
particularly worrying because the largest share of
grant monies is disbursed through it and technology-
intensive firms take their chances along with firms
in traditional industries that lay claim to
assistance because of their employment creation in
more depressed regions.

Patterns of Bias in IRDP Impact. DRIE was formed in
1983 with tasks that are an amalgam of (1) concerns
for regional development, primarily outside the
metropolitan industrial regions, and (2) industrial
support responsibilities, particularly assistance
for expansion, innovation and technological
upgrading. The regional tasks devolved from the
former Department of Regional Economic Expansion and
the technological support responsibilities were
previously specified in the Enterprise Development
Program (EDP) of the former Department of Industry,
Trade and Commerce. Reactions to IRDP generally

recognise that regional and technological development interests are rarely complementary. IRDP is a
broad spectrum industrial support programme designed
to combat regional disparities and its assistance is
not specialised among the components of the product
cycle; it is concerned with all phases from
innovation to marketing, and from there to
restructuring. Furthermore, it is open to
applications from large firms, even foreign
subsidiaries, and has proven itself sensitive to
their seeking 'modernisation/expansion' assistance.
In 1984-85, for example, 34 per cent of IRDP
assistance ($142 million) was disbursed in 10 lots of
$5 million-or-more only two of which were
specifically concerned with innovation. Two very
large grants were $60.5 million to American Motors
Corporation and $22 million to General Motors of
Canada: both these auto projects were for
modernisation/expansion. With very large grants like
these influencing the statistics it is difficult to
interpret the industrial distribution of assistance
but low rankings for medium technology industries
indicate that an interest in plant expansions and
jobs compromises the IRDP support for high-
technology industry (Table 6.20). This is further
illustrated when the grants exceeding $5 million are
subtracted from industry totals: only then does
electrical equipment attain first rank followed by
other industries and machinery, a pattern which is a
more reasonable path to technological development.
    Although technology intensive industry is
concentrated in a few metropolitan areas and although
these are fertile locations for new enterprises, the
IRDP is biased against metropolitan areas through a
four-tier system. Tier 1 areas, which contain the
most developed parts of Canada and about 50 per cent
of the population, generate a proportionally lower
level of financial assistance. As a group, Tier 1
received only 42 per cent of the total assistance and
this again reflects the federal political compromise
between technological development and the creation
of employment opportunities. The more defensible
approach, given Canada's technological gap, would be
to support strengths in Canada's technology-based
industry and this would mean that regions like
Toronto would receive assistance at least in line
with their capacity to generate technology-based
growth. In practice firms in the Toronto region
received $21.3 million in assistance in 1983-84, not
counting the $60.5 million grant to AMC, and this
support was only 6 per cent of the adjusted Canadian

Table 6.20: Canada: IRDP Assistance 1984-85 Sectoral Ranks

	Number of Firms Assisted			$ million of Assistance	
1.	Metal fabricating	166	1.	Transport equipment	99.5
2.	Food	161	2.	Electrical equipment	84.6
3.	Machinery	149	3.	Other	32.5
4.	Other	145	4.	Machinery	22.2
5.	Electrical equipment	141	5.	Food	22.0
6.	Tourism	124	6.	Primary metals	21.3
7.	Wood inds.	108	7.	Wood inds.	21.0
	Total	1464		Total	$418.1

Source: Canada, Regional Industrial Expansion, Industrial and Regional Development Program, Annual Report 1984-1985.

total despite Toronto's 12.4 per cent population!

Policy Changes. The replacement of the federal Liberal with a Conservative government has initiated policy changes both of an organisational type and in the financial base of support for high-technology industry:

1. Assistance for small firms that wish to raise equity has been provided through a variety of mechanisms - a wider variety of venture capital pools that derive their input from individuals earning tax credits, and liberation of pension funds to invest overseas in return for a 3:1 leverage of investments in SMEs in Canada. It is thought that up to $5 billion could become available if there are SMEs to absorb finance.
2. The refund of tax credits to firms undertaking R&D is increased for small firms and only Canadian-controlled firms will get these R&D tax breaks.
3. The definition of R&D has been broadened to include personnel and equipment partly involved in R&D, and the experimental development of new products and processes.
4. To rationalise support for R&D the proposal has been made within the government to combine the IRAP and PILP programmes of NRC with IRDP. While this echoes a recommendation made in the Wright Report (1984) it remains to be seen whether DRIE or NRC will be the controlling agency.

These changes in the policy environment provide a fiscal push and venture capital access for small high-technology business, they will possibly reduce the maze of programme specifications, and they introduce a sensible concept of R&D. But they do not address the locational bias in IRDP assistance and do not advance any concept of active policy delivery which will be needed if the flow of venture capital is to be used by small business.

Conclusions

By use of a variety of data it has been possible to develop clear evidence on Canada's problem of technological development and on the negative aspects of the high level of foreign control, and to generate locational information that shows well defined patterns of growth of technology intensive manufacturing. A major weakness in the basic data file, however, is the omission of computer software producers from official industry-based statistical enquiries. The resolution of the problem is to compile data from directories and other records so that the scale and location of the software industry can be evaluated. Once this task is completed serious enquiries can begin among computer software and other business service firms. It is particularly important that the functional linkages within Toronto's economy are understood otherwise the importance of technology intensive services for medium and low technology firms, that survive through the quality of their innovation, will not be fully appraised. Likewise, it is necessary to document the efficacy of supply-side initiatives taken by the provincial government. The omission of Toronto from the locations of Ontario's technology centres represents 'opportunity' for other centres but it also generates an eccentric penalty for the firms located in the Toronto region. Finally, the productivity of Canadian controlled firms in spinning-off new enterprises urgently demands research. Currently the evidence is fragmentary and in growth locations there is important evidence to be uncovered on the corporate and institutional origins of new firms.

Too few of Canada's large technology-based firms are under domestic control. It is likely, therefore, that there are reduced corporate sources of new enterprises. Foreign firms have not yet widely adopted the world product mandate form of Canadian production and, therefore, they have few activities concerned with product development that could become the basis of new enterprises. Nevertheless, the trend to rationalisation of U.S. firms is increasing and the traditional supply of mature products for the Canadian market has given way to technology-based product lines in a number of cases. It is evident that the further corporate elimination of the plants of subsidiaries that produce mature products will occur unless an R&D operating model of greater intensity and export orientation is adopted and thus there is a perceived need for Canadian policy to assist the transformation, and thus survival of jobs in foreign controlled firms.

Largely, Canadian policies do not differentiate between foreign and domestically-controlled firms though the tax credit instance where there will be differences in treatment has been noted. Even the Foreign Investment Review Agency has been disbanded.

Nevertheless, chances for major improvements in
Canada's industrial performance probably lie with
the development of domestically controlled SMEs even
though at their present size they are not large
employers.
    Canadian governments have built an industrial
policy environment containing two elements.

1.   a technological component that assists R&D and
      high-technology industry and
2.   a regional development thrust that addresses
      disparities in regional development through
      assistance for industrial expansion.

For a long time these elements were separated: the
regional development issue was pursued partly
through assistance to firms that would generate
employment expansion and partly through infrastructural investments. Technological development was
addressed by means of a variety of mechanisms to
encourage the growth of research intensive firms and
the use of technology. Recently, at the federal
level, these two elements have been unsuccessfully
confounded within a grant programme. This
illustrates the fact that Canadian analysts and
Canadian policy makers have been quite unclear about
the significance of locational factors influencing
technological performance and the importance of
agglomeration economies in the generation and growth
of high-technology enterprises has never been
adequately recognised. It might be thought that the
publicity surrounding the rise of the Ottawa area as
a technology concentration would have sharpened
understanding of the conditions necessary for the
emergence and growth of high-technology industry. In
Ottawa's case the dependence on government
contracts, a highly skilled technical labour pool,
and spin-offs from existing enterprises are all
important in producing Ottawa's modest but
surprisingly localised growth. In fact, this
understanding has not transpired and Ottawa's growth
may have encouraged the distribution of assistance in
such a way that other smaller versions of Silicon
Valley North might be encouraged. A more cynical view
would, of course, hold that the failure of policy to
recognise locational principles and to attain
geographic coherence is a logical result of the
political process.
    There is, however, an increasing sense among
high-technology industry observers that Ottawa has
experienced its period of expansion and that it is

losing increasingly to other locations, Toronto in particular, because of the location of large customers. The evidence points to several communities in the Toronto region actively generating high-technology firms and accommodating some of the larger companies in the country. Metropolitan Toronto itself and Mississauga and Markham are important components of the high-technology structure of the Toronto region. Their growth reflects the region's sectoral diversity and its infrastructural depth - its universities, its headquarters functions, its cultural amenities and its centrality to the Canadian industrial economy.

Toronto was well served under the older generation of technology support programmes; now, however, the administration of direct technological development assistance is biased against equitable support for technology-based firms in the Toronto region. But it is not in the interest of Canada to proceed in this manner when the technology gap is large. Agglomeration advantages could be taken as the locational signal for allocating greater assistance in the hope of generating greater impact. In general, what is required of policies is that they should operate through, not against, existing locational strengths.

Acknowledgement

The research assistance of Lance Alexander in collecting and processsing data from the Census of Manufacturers, and in combing industrial directories is gratefully recorded.

References

Britton, John N.H. (1979) 'Canada's Industrial Performance and Prospects Under Free Trade' Canadian Geographer, XXI, 351-71

Britton, John N.H. (1980) 'Industrial Dependence and Technological Underdevelopment: Canadian Consequences of Foreign Direct Investment' Regional Studies, 14, 181-99

Britton, John N.H. (1985) 'Research and Development in the Canadian Economy: Sectoral, Ownership, Locational, and Policy Issues' in R. Oakey and A. Thwaites (eds) Technological Change and Regional Economic Development Frances Pinter, London

Britton, John N.H. and Gertler, Meric S. (1986) 'Locational Perspectives on Policies for Innovation' in Jerry Dermer (ed.) Competitiveness Through

<u>Technology: What Business Needs From Government</u>, Lexington Books

Britton, John N.H. and Gilmour, James M. (1978) <u>The Weakest Link: A Technological Perspective on Canadian Industrial Underdevelopment</u> Science Council of Canada, Ottawa, Background Study 43

Cohen, J.S., Rubin, J. and Saunders, R.S. (1984) 'Chasing the Bandwagon: Government Policy for the Electronics Industry' <u>Canadian Public Policy</u> 10, 25-34

Daly, Donald J. (1979) 'Weak Links in the "Weakest" Link' <u>Canadian Public Policy</u>, V, 307-17

Economic Council of Canada (1975) <u>Looking Outward: A New Trade Strategy for Canada</u>, Ottawa

Economic Council of Canada (1983) <u>The Bottom Line: Technology, Trade and Income Growth</u>, Ottawa

Federal Business Development Bank (1984) <u>Assistance to Business in Canada</u>

Layne, Donald (1982) 'The Performance of Small Manufacturing Units' Conference papers: International Council for Small Business, Edmonton, 147-57

MOSST (1978) <u>Directory of Scientific and Technological Capabilities in Canadian Industry 1977</u>, Ministry of State, Science and Technology Canada

Naylor, R.T. (1975) <u>The History of Canadian Business 1967-1914</u>, 2 volumes, Toronto: Lorimer

Oakey, R.P. (1984) 'Innovation and Regional Growth in Small High Technology Firms: Evidence From Britain and the USA' <u>Regional Studies</u>, 18, 237-51

Ondrack, D.A. (1980) <u>Innovation and Performance of Small and Medium Firms: A Re-Analysis of Data on a Sample of Nineteen Small and Medium Firms in the Machinery Industry</u>, Industry, Trade and Commerce, Technology Branch, Ottawa

Rothwell, Roy (1984) 'The Role of Small Firms in the Emergence of New Technologies': <u>OMEGA</u>, 12, 19-29

Rugman, Alan M. (1981) 'Research and Development by Multinational and Domestic Firms in Canada' <u>Canadian Public Policy</u>, VII, 604-16

Safarian, A.E. (1979) 'Foreign Ownership and Industrial Behaviour: A Comment on the Weakest Link' <u>Canadian Public Policy</u>, V, 318-35

Saxenian, Anna Lee (1984) 'The Urban Contradictions of Silicon Valley: Regional Growth and the Restructuring of the Semiconductor Industry', in L. Sawers and W.K. Tabb (eds) <u>Sunbelt/Snowbelt: Urban Development and Regional Restructuring</u>, Oxford University Press, New York, 163-97

Segal, N.S. (1983) 'Universities and Advanced Technology New Firms in Great Britain' Conference paper: International Workshop on the Future of Industrial Liaison, Technische Universität, Berlin, Nov. 1983

Science Council of Canada (1984) <u>Canadian Industrial Development - Some Policy Directions</u> Ottawa: Report No. 37

Steed, Guy P.F. (1982) <u>Threshold Firms: Backing Canada's Winners</u> Science Council of Canada, Ottawa, Background Study 48

Steed, Guy P.F. and de Genova, Don (1983) 'Ottawa's Technology-Oriented Complex' <u>Canadian Geographer</u>, XXVII, 263-78

Storey, D.J. (1983) 'Job Accounts and Firm Size' <u>Area</u>, 15, 231-7

Wright Report (1984) <u>Report of the Task Force on Federal Policies and Programs for Technological Development</u> Ottawa: Ministry of State, Science and Technology

Chapter Seven

THE LOCATION OF HIGH TECHNOLOGY INDUSTRIES IN FRANCE

Claude Pottier (CNRS, University of Paris)

Two Approaches to High Technology Industries

The most widely accepted theory among modern-day economists concerning the development and location of high technology industry is the product cycle thesis: new branches of industry sprout up in large urban centres as a result both of a technically highly skilled workforce and the size of the market, and then spread out into the peripheral areas where a less skilled workforce work on production lines using fixed techniques. However, this vision of things is being challenged by many authors who see a new trend towards research and development (R&D) decentralisation (for France, see Planque, 1983), as well as a rapid growth in local centres helping industrial firms to innovate and modernise.

The process of industrial decentralisation which took place in the Paris area between the mid-1950s and the mid-1970s has been adequately covered in the analysis of the product cycle. However, with the development of the crisis from the mid-1970s on, a reorganisation of the production system would appear to have been taking place under the double impulse of the opening up of national markets and the development of the technological revolution. The distinctive roles of these two factors do not seem to have been sufficiently underlined in the literature. This may be done as follows: firstly, the opening up of national economies would provoke market instability which in its turn calls for greater production flexibility. This would mean smaller production units whether it be independent companies or companies within a group but each with increasing autonomy. At the same time, links between research and production would become closer and more permanent. Secondly, the micro-electronic revolution would allow production to become computerised and

automated, meeting the call for more flexible
production.

This development may have several spatial
consequences. The closer links between R&D and
production may strengthen the supremacy of large
urban centres where research units are concentrated.
On the other hand, the need for permanent innovation,
improved circulation of technical information and
the geographical mobility of engineers may bring
about the decentralisation of the most technical
jobs. To see how this opposition can be overcome, it
would be useful to distinguish two things at the
outset and this will serve as the basic hypothesis of
this article. Firstly, there are major innovations
such as the appearance on the market of new products
or the creation of new industrial branches. In these
cases the product cycle thesis is probably still
valid. But there has recently arisen a need for
innovation to spread out, to adapt products to the
growth in demand in an increasingly international and
unstable market, as well as to adapt new
technologies, in particular the development of
automation. These minor but permanent innovations
involve a certain amount of decentralisation. This
leads us to a double approach and a double definition
of high technology industry:

1.   A sectorial approach: high technology industry
      is a sector of production where the level of
      technical skill is high as shown by the
      technologically skilled workforce or the
      amounts of money spent on R&D.
2.   An overall approach: given the generic nature of
      new technology and its impact on industry as a
      whole, we can no longer think in purely
      sectorial terms.

True, the regions have been greatly influenced
by sectorial specialisations but this is precisely
what is changing. In every sector of manufacturing a
'mediate work' (as Mahieu puts it) is developing,
that is, both the preparatory work upside of the
manufacturing process and the intellectual component
constituted by the work of an operator calling upon
his powers of abstraction and logic (see Delpierre,
1985). From this point of view, high technology
industry is defined as design and final technical
adjustment prior to production. This stage may be
analysed with the help of two indicators:

1.   The degree of technical skill of production

labour in manufacturing, shown by the proportion of production technicians and engineers within the production staff as a whole (made up of production technicians and engineers plus foremen and workers). Administrative and commercial managers as well as office staff are not taken into consideration.

2. The size of industrial R&D revealed by the proportion of personnel assigned to it within the manufacturing industry workforce as a whole. They are mainly engineers and technicians but some labourers are also involved.

But one may also ask why the study should be limited to manufacturing when the technological revolution is developing the service sector too. It must be pointed out that our two indicators measure a part of the service activity within the manufacturing sector, which we may call 'industrial service sector', i.e. services directly oriented towards production of goods. Besides, there exists, in the French industrial classification, two sectors of technical assistance to firms: 'technical analysis' and 'data processing and organisation analysis'. It would therefore be interesting to study the evolution of their manpower from a regional point of view. Unfortunately, the number of replies to the Structure of Employment Inquiry (carried out by the INSEE) into these sectors is both too small (some two-thirds) and too varied in time. The poor response can be explained by the great number of industrial firms with below ten employees, which were not included in the inquiry. As we also have reason to believe that the distribution of these firms is not uniform according to region this data must be rejected.

Location Hypotheses

As has been stated, any discussion of the location of high technology industry opposes the idea of centralised development to that of decentralised development. On the one hand, the importance of research and the highly technical nature of work in the large urban centres should be an even greater asset in the technological revolution currently taking place. On the other hand, the demand for permanent and generalised innovation does favour decentralisation. At this point, it is worthwhile noting that development centres can change. As Aydalot (1985) clearly states, centrality is not to be confused with size of towns. Any large town is not destined to be or remain a development centre. Towns with a large service sector even before the industrial revolution of the nineteenth century can especially find themselves favourably placed in the 'post-industrial era'. This much-travelled expression must not be taken to mean the weakening of manufacturing within the economy but rather as the growth of intellectual activity linked to manufacturing. Studies must therefore be carried out not only per size of town but also taking into account the nature of the towns or regions and thus examining the size of the 'industrial service sector'.

Priority will be given to the regions and they will be distinguished according to the size of their industrial service sector and the degree of diversification of manufacturing industries, diversified manufacturing being on the face of it a factor favourable to the growth of new industry. Another hypothesis is that some specialisations may be advantageous because their technology is complementary with that of the new activities. However, regions specialised in declining industries are at a disadvantage when it comes to renewing their industry, not only because of this specialisation but also because they are branded socially with an industrialisation out of its depth. Some authors insist on this view of things: The size of salaried workforce in the large industrial plants of these regions is matched against the dynamism of small industrial firms in other regions. Labour is skilled but with skills that are no longer in demand. There exist various forms of social organisation, mental outlook, working-class culture and even industrial landscape all of which hold a critical place in modern society. Here, we must tread carefully: some forms of social organisation are clearly obsolete but it is precisely here that the governing classes must take great care to show that they are indeed beyond salvation. Such is the technologist ideology behind the study of industrial mutations, emphasising the conflict of outlooks, leaving apart the opposition of classes.

At the other end of the scale, we must consider the importance of 'the quality of life' in the location of high technology industry. Given that this industry usually does not suffer the constraints of transportation costs, location criteria may be linked to the desires of management looking to flee the factory chimneys and come closer to nature, sunshine and leisure activities. As a final point, the links between high technology industry location and certain signs of regional dynamism must be analysed, e.g. the rate at which new industrial firms are being set up and the level of immigration. To test all these hypotheses, I have formed economic areas covering the French regions, based on the criteria established above.

The Overall Approach

Before proceeding, we may examine the level of technical skill, within the manufacturing workforce, per size of urban unit.

Table 7.1: Level of Technical Skill (a) in Manufacturing Workforce in French Urban Units of over 10,000 Inhabitants

Agglomerations (no. of inhabitants)	1975	1980	Evolution indexes (1975 = 100)
10 - 20,000	6.4	8.2	128
20 - 50,000	7.0	8.6	123
50 - 100,000	8.4	11.3	135
100 - 200,000	9.8	12.8	131
200,000	11.1	14.6	132
Paris agglomeration	21.2	23.2	109

Note: (a) Proportion of production technicians and engineers within the manufacturing industry's production staff as a whole.

Source: INSEE, Inquiry into Employment Structure.

Two remarks can be made from Table 7.1:

1.   The level of technical skill increases steadily with size of urban unit.
2.   The hierarchy is not challenged between 1975 and 1980 but the supremacy of the Paris agglomeration diminishes and disparities widen at 50,000 inhabitants.

In her study into the development of interurban disparities in France, Saint-Julien (1984) compares those found throughout industry and those found in high technology industry as a whole. She finds that in the case of the latter the level of labour technical skill increases at a much greater rate as

196

the size of urban units increases. She goes on to say that the relative erosion of Paris' domination is even less marked in high technology industry and concludes that trends in location of the latter are strengthening the traditional differentiations of work skill according to size of town.

With the recent development of periurban zones, the study must now be extended to areas beyond urban unit limits. The French regions are therefore grouped according to the criteria established above and which are based on the hypotheses put forward for the location of high technology industries. Table 7.2 and the map, which follows it, show this regional grouping with, for each group, the degree and age of industrialisation, the impact of the 1960s decentralisation process, the proportion of staff in large manufacturing plants, the level of worker qualification, the degree of urbanisation and the rate at which agricultural population dropped from 1970 to 1980. The data is generally that for 1975 which marks the beginning of the period under analysis.

From among the regions of early industrialisa-tion, four groups may be distinguished.

1. The Paris region (Ile-de-France): the nucleus of the national economy, the headquarters of industry where the industrial service sector is concentrated. Manufacturing developed early with high growth between the wars.
2. Rhône-Alpes: another centre of diversified manufacturing industry, both early and recent, a large service sector. This region cannot be classified along with the Paris region such is the domination of the latter. From the facts, there is every reason to believe that these two regions will be conducive to the development of high technology industry with their large service sectors and the diversity of their industry.
3. The Nord-Pas-de-Calais and the Lorraine, unlike the first two, specialise in industries which have been in decline since before 1975 (coal, iron and steel, textiles). The proportion of large industrial plants is high. Worker skill is high but results from know-how accumulated over generations rather than from technical knowledge adapted to current needs.
4. Lastly, the other regions of early industriali-sation can all be grouped together into one. They are, on the one hand, regions bordering on

Table 7.2: Main Indicators of Manufacturing Industry for French Region Groups at the Beginning of Period of Analysis

	Degree of ind. a	Age of Indust. b	Indust. decent. c	Large Plants d	Skilled labour e	Degree of urban. f	Agric. Emigra. g
Paris region	28	8	0	60	121	99	0.1
Rhone Alpes	36	36	2	65	94	88	5.0
Nord-Lorraine	41	-8	3	81	116	94	2.9
Other regions of early industrialisation	37	28	10	76	81	82	4.0
West	26	32	16	69	97	70	8.3
South-West	21	13	4	56	129	72	9.8
Provence-C.d'Azur	19	19	3	57	170	95	2.5
France	27	17	6	68	99	83	4.9

Notes:

a   Proportion of employment in manufacturing in 1975 (%).
b   Variation in manufacturing employment from 1954 to 1975 proportionately to manufacturing employment in 1975 (%).
c   Number of decentralised manufacturing plants from 1951 to 1975 per 10,000 jobs in manufacturing in 1975.
d   Proportion of manpower in manufacturing plants with over 100 employees in 1979 (%).
e   Ratio between number of skilled and unskilled workers in 1975 (%).
f   Proportion of population living in urban areas (Z.P.I.U) in 1975 (%).
g   Drop in agricultural population from 1970 to 1980 in proportion to working population in 1975 (%).

Sources: INSEE, AUREG (c.), Ministry of Agriculture (g.).

Figure 7.1: French Region Groups and their Industrial Employment in 1982

the Paris region where the decentralisation of Parisian industry added to early established industry, namely Haute Normandie, Picardie, Champagne-Ardennes, Bourgogne and, on the other hand, two more peripheral regions, Franche-Comté and Alsace, where again large manufacturing plants dominate but where labourer skill is low.

Among the least industrialised regions three criteria lead to differentiation: the degree of Parisian decentralisation, the level of urbanisation and the worker qualification.

1. The regions of the West (Basse Normandie, Centre, Pays de la Loire, Bretagne and Poitou-Charente) are regions of low urbanisation where manufacturing is for the most part the direct result of Parisian decentralisation. Worker qualification is not far from the national average but is low in decentralised manufacturing plants, the latter employing a workforce coming from agriculture.
2. The three regions of the South-West (Aquitaine, Midi-Pyrénées and Languedoc-Roussillon) are regions of low urbanisation where the agricultural population dropped appreciably from 1970 to 1980. However, this population did not migrate into manufacturing in massive numbers, as was the case in the West, since decentralisation was low here. Labourer qualification is high, thanks mostly to the large aeronautics industry, itself linked to a decentralisation process which took place well before the 1960s, before the war in fact.
3. The Provence-Côte d'Azur region is rather specific as industrialisation is low but urbanisation is high. The level of industrial service activity is high as is labourer qualification.

As far as these three types of region are concerned, it is important to know whether their generally dynamic nature, taking into account the relative development of manufacturing employment, the number of new industrial firms and immigration (see Table 7.3), is accompanied by a growth in high technology industry. On the face of it, this should be the case for Provence-Côte d'Azur which boasts two important advantages: a highly developed service sector and a pleasant environment. The western regions, however, present a problem: was the industrial decentralisation of the 1960s, which mainly involved the execution of pre-established jobs and unskilled labour, great enough to create a real development momentum and a significant growth in high technology industry?

Table 7.3: Indicators of Dynamism of French Region Groups, 1975-1982

	Variations in manuf. employ. 1975-1982 (%)	Rate of new ind. firms 1974-1982 a	Rate of net immigr. 1975-1982 b
Paris region	-16	26[c]	-3.0
Rhône-Alpes	-10	17	+1.1
Nord-Lorraine	-17	12	-3.8
Other regions of early indust- rialisation	-8	13	-0.7
West	0	21	+1.7
South-West	-5	34	+4.5
Provence-Côte d'Azur	-8	57	+6.6

Notes:
a. Number of industrial firms set up from 1974 to 1982 for every 1,000 industrial jobs in 1975.
b. Migration balance from 1975 to 1982 in proportion to the population in 1975 (in %).
c. A certain number of firms in the provinces have their registered offices in the Paris region.

Sources: INSEE and CEPME.

Lastly, three regions where there are problems of access, Limousin, Auvergne and Corse, are not taken into account in this regional grouping. For Limousin and Corse, industrialisation is so low that variation indexes are of little significance. As for data concerning Auvergne, it is highly dependent upon the performance of one large company, Michelin.

We may now examine the level and evolution, within these regions, of our two overall indicators showing the dimension of high technology industry, i.e. the level of technical skill in production labour and the place of R&D employment in manufacturing.

Table 7.4 shows the state of labour technical

skill in manufacturing production. The Paris region is far and away ahead of the field, followed by Provence - Côte d'Azur. Bringing up the rear are Nord - Pas-de-Calais and Lorraine, the 'other regions of early industrialisation', and the West, all of which have low coefficients. Evolution over the period from 1975 to 1981 shows great stability among disparities. Taking into account statistical inexactitudes, the only noteworthy development has taken place in the South-West where the coefficient is rising faster than the national average. This stability is remarkable as manufacturing employment developed in a very divergent way within the regions from 1975 to 1982 (see Table 7.3).

Table 7.4: Degree of Technical Skill (a) in Production Labour within the French Region Groups

	1975	1981	Variation 1975 = 100
Paris region	19.7	25.0	127
Rhône-Alpes	9.5	12.0	126
Nord-Lorraine	6.6	8.5	129
Other regions of early industrialisation	7.0	8.8	126
West	6.5	8.1	125
South-West	9.0	11.9	132
Provence-Côte d'Azur	13.2	16.0	121
France	9.8	12.2	125

Note: a. Proportion of production technicians and engineers employed in manufacturing production.

Source: INSEE (Inquiry into Employment Structure)

Table 7.5 shows the size and evolution of R&D within manufacturing:

1. The regional hierarchy of R&D bears a great resemblance to that of production labour technical skill in that the Paris region is again a long way ahead followed by Provence - Côte d'Azur. The spatial distribution of employees shows that these two regions can not be compared: 55 per cent of R&D staff in the Paris region for 5 per cent in Provence - Côte d'Azur.

2. However, variation in the importance of R&D in manufacturing employment from 1975 to 1982 is

Table 7.5: Evolution of R&D Employment in Manufacturing within the French Region Groups

	R&D empl. in % of manuf. empl. 1975	Var. in R&D empl. 1975-1982 %	Var. in manuf. empl. 1975-1982 %	Var. in R&D empl. within man. empl. 1975=100	R&D empl. 1982 % of France
Paris region	5.18	+9	-16	129	55.2
Rhone-Alpes	1.72	+8	-10	121	9.9
Nord-Lorraine	0.47	+2	-17	123	3.3
Other regions of early industrialisation	0.78	+7	-8	115	7.9
West	0.56	+41	0	139	6.1
South-West	1.97	+7	-5	112	7.9
Provence – Cote d'Azur	2.41	+14	-8	123	5.3
France	1.96	+9.5	-9.5	121	100.0

Source: Ministry of Research (DGRT)

203

far less uniform. The Paris Region is strengthening its domination as a result of a fall in aggregate manufacturing employment whereas the number of persons employed in R&D is rising more or less with the average for the regions. In the western regions, manufacturing employment is stagnant but the number of persons employed in R&D is rising sharply. On the contrary, in the South-West and 'other regions of early industrialisation', R&D personnel is rising slowly, so much so in the South-West as to find itself in direct contradiction with the favourable evolution seen in the production technicity indicator. As for the growth in the relative importance of R&D in Nord - Pas-de-Calais and Lorraine, this is the result of a sharp fall in manufacturing employment with a rise in the number of R&D personnel well below the national average.

Whereas the production technicity indicator led us to believe that the regional hierarchy was extremely stable, R&D development within manufacturing leads us to the following conclusions:

1. R&D is spreading in the West, i.e. in regions where it used to be low. Specialised though these regions may be in traditional labour intensive industries, R&D is growing fast, much faster in fact than anywhere else, both in incremental rate and in proportion of manufacturing employment as a whole.

2. On the other hand, Nord - Pas-de-Calais and Lorraine are by far the regions where R&D development is slowest even though it was already very low in 1975. This would seem to prove the hypothesis, put forward in this article, with reserve, of an overall inadaptability of these regions with old traditional industries. But it must be concluded from the clearly more favourable evolution of R&D in the 'other regions of early industrialisation' that poor sectorial specialisation is of great importance. This will have to be taken up again when the sectorial approach is studied later.

3. The spread of R&D in some regions is in no way a challenge to the Paris region. In proportion of manufacturing employment as a whole, R&D is growing even faster in the Paris Region than the national average, this because of the drop in

manual labour within the region.
4.   Finally, the example of Provence – Côte d'Azur shows that the model of heliotropic location is to a certain extent valid. Here is a region where R&D, already high, is rising faster than the national average. As technical constraints concerning the location of research units are insignificant, factors of attraction for engineers – sunshine, scenic environment, leisure activities, etc. – come into full play. But a good telecommunication network as well as easy access to international airports are necessary, and these can be found in Marseille and Nice. The importance of Provence – Côte d'Azur must not, however, be exaggerated. As Saint-Julien (1984, p.27) puts it, 'It would be derisory to have visions of a new California in this region'. Moreover, this growth in R&D may not necessarily be linked to the dynamism indicators seen for this region where the level of immigration and the number of new industrial firms from 1975 to 1982 were the highest in France. Comparisons carried out within the 21 regions between on the one hand, the level of production technical skill in 1975, growth in the degree of technical skill from 1975 to 1981 and growth in R&D personnel from 1975 to 1982 and, on the other hand, the rate at which new industrial firms were set up from 1974 to 1982 and the net rate of immigration from 1975 to 1982 show no significant correlation.

The Sectorial Approach
High technology sectors can be identified by means of the two indicators previously used, i.e. the degree of technical skill in production labour and the size of R&D, the latter indicator being based on data concerning the number of persons employed or the spending involved. Rather than going blindly into a general analysis of the high technology sectors as a whole, it would be more interesting, especially after our overall approach, to study specific sectors, chosen on three criteria. The first is the age of the sector: some sectors have been exploiting high technology for a long time now and show location structures which go back many years. The second is the degree to which the sector is open to world markets. Finally, some sectors are typically located in some of the groups of regions studied in the previous section.

Four sectors will come under analysis (briefly, for lack of space).

1. Microelectronics, a rather new sector in the international division of labour.
2. The telephone industry, which is undergoing a profound technological transformation. The example taken is Brittany, a particularly interesting case of a high technology industry in an area of low industrialisation.
3. Robotics, a very new sector, which will enable us to test the sectorial complementarity hypothesis in certain regions of early industrialisation,
4. Aeronautics, an early sector which has been decentralised for some time.

The Micro-Electronics Industry. The current siting of micro-electronics firms in France is the result of two distinct processes (see Pottier and Touati 1981, Pottier, 1983):

1. American firms established in France from the mid-1960s on, without any noteworthy change in the specialisation of these establishments within the American companies international division of labour.
2. A process of decentralisation away from the Paris region by French firms and by RTC, a subsidiary of the Philips group, from the mid-1950s on, followed by the transfer of a large part of assembly work - from the mid-1970s on - to third-world countries.

Within the American firms, we must different-iate between those for which output of semi-conductors forms an integral part of other activities and those whose output is directly market-oriented. For the latter, the international division of labour was straightforward: fundamental research remained in the United States while the greater part of assembly, as far as mass production was concerned, was immediately located in South-East Asia, French factories housing research with a more limited scope and production of a highly technical nature. For this, in 1966, Motorola set up in Toulouse, a city with a large university specialising in electronics and with an international airport. Texas Instruments, set up in Nice from 1961, disposes of a much more limited university environment but Nice does have an international airport and also an excellent telecommunications system. Here I second the remarks made by H. Nishioka (1983), concerning Japan, when he states that the proximity of airports is an essential factor in the siting of plants producting integrated circuits and, in a wider sense, in the location of many high technology industries where transport costs represent a small part of final cost price. What goes for the island of Kyushu goes even more for American companies ensconced in the international division of labour. Toulouse's pleasant environment and especially Nice's was also of some importance, all the more so as engineers had to be brought over from the United States. The setting up of two other American companies, in joint venture with French firms, was carried out - much more recently, in 1977 - along the same lines, namely by National Semiconductor near Marseille and Harris in Nantes. Pressure was also applied by the authorities to have the plants located in these areas of high unemployment.

Things are different when it comes to firms producing semiconductors for their own use, IBM and ITT. IBM production has been carried out in its leader unit in France, in the Paris region, since 1955, the unit having worldwide responsibility for a certain type of integrated circuit in conjunction with a laboratory situated in the United States. Assembly is carried out on the spot and seems to be relatively highly automated. The number of highly qualified personnel justifies location in the Paris area. On the other hand, ITT's plant is situated in an average-sized town in Alsace. The relevant criterion for this siting is proximity to West Germany where the company has its semiconductor production centre. As in the previous case assembly is carried out in France (as well as in West Germany and the United Kingdom but not in South-East Asia) because of the high degree of automation.

Decentralisation from the Paris region of French companies (notably Thomson) and of RTC, a subsidiary of Philips, began in the mid-1950s. Given the importance of assembly work in the case of 'discrete' semiconductors (such as the transistor), before integrated circuits came on to the market, the companies set up in the provinces to take advantage of a workforce paid 30 per cent less than in Paris. The companies set up plants in small or average-sized towns in the West, which happened to be places of settlement after the rural exodus, but also in university towns which were more conducive to the hiring of qualified staff: Caen and especially Grenoble. In fact, the growth in integrated circuit production widened the gulf between large university towns and average-sized towns. Philips still carried out fundamental research in the Netherlands and, to a lesser extent, in the Paris region but their Caen plant became world leader in the design and production of a certain type of circuit. As for Thomson, they transferred their entire research centre to Grenoble, which then became the centre of micro-electronics in France boasting Thomson's two production units, two large state-run research centres, the university and a number of higher university engineering and technical institutes. In the Grenoble and Caen production plants, production manpower technical skill is high.

On the other hand, the plants in small or average-sized towns in the West specialised in the production of 'discrete' semi-conductors or in assembly of all semi-conductors and this specialisation was challenged when control of semi-conductor production became such that it was possible to transfer assembly to third-world countries around the beginning of the 1970s. Another reason for this transfer was a change in strategy on the part of European firms towards their American and Japanese competitors. Whereas previously they had had to produce a small number of all the different kinds of components with the help of government authorities, there was now a stricter separation between, on the one hand, research and the manufacture of high technology products and, on the other, mass production, under foreign licence, of standard products, for which the drop in assembly costs was particularly advantageous. As a result of this development there was a drop in the sector's workforce but an increase in production workforce technical skill within all plants, including those in the West.

The Telephone Industry in Brittany. The great technological change in the telephone industry was the switch from electromechanical to electronic commutation. The effects of this change, brought about after many years of research, were felt in concrete terms from 1977-78 onwards in France. At the very beginning of the 1970s, a French company, CIT Alcatel (CGE group), has developed a sophisticated commutation process but only controlled 30 per cent of the French market compared with 60 per cent controlled by the subsidiaries of foreign companies, ITT and Ericsson, who were technologically far less advanced. In 1976, Thomson broke onto the market by taking over one of the ITT subsidiaries (LMT) and the Ericsson subsidiary. In 1982 both French groups were nationalised and the state bought out ITT's remaining subsidiary, CGCT, which meant that state-owned companies controlled the entire public commutation market, 70 per cent of the transmissions market and over 50 per cent of private telephony. In 1983, the CGE group took over Thomson's telephone activities and thus controlled the whole market.

From a spatial point of view, the telephone industry offers the most striking example of decentralisation in the 1960s. With a large-scale transfer of activity from the Paris region to Brittany, the least industrialised and most peripheral region in the West. This process was given a boost when the State decentralised research centres and certain 'grandes écoles' (higher university colleges). Scientific nuclei were thus formed, attracting research teams from large companies which were also able to transfer production units to employ local cheap labour. State subsidies were a further important factor contributing to this migration.

First of all, how many jobs were created, what was the degree of qualification required? In 1975, about 9,000 people were employed in the telephone industry in Brittany compared with 75,000 nationwide and 35,000 in the Paris region. These 9,000 people represented 4.9 per cent of Brittany's manufacturing employment whereas nationwide, the telephone industry represented only 1.2 per cent of manufacturing employment, giving Brittany a very high location quotient, slightly over 4. Table 7.6 gives indicators of job qualification. It shows that the technical level of production manpower was much lower in Brittany than in the Paris region, almost four times as low for technicians, more than nine times as low for engineers. The level of worker qualification was also much lower. The spread of this high technology industry to Brittany thus went along with the transfer of technically skilled labour but the technical disparity with the Paris region was still very high in 1975 when decentralisation was coming to an end.

Secondly, was Brittany's telephone industry able to develop at all independently after 1975? To be more precise, if this ability to develop did exist, we should talk of an ability to reconvert as switching from electromechanical to electronic

Table 7.6: Qualification Indicators in the Telephone Industry in Brittany and the Paris Region

	Proportion of engineers in prod. jobs (%)		Proportion of technicians in prod. jobs (%)		Proportion of skilled workers (a) among workers as a whole (%)	
	1975	1981	1975	1981	1975	1981
Paris region	13.1	17.8	29.2	37.8	74.1	79.5
Brittany	1.4	7.8	7.5	18.5	20.1	35.9
Paris reg./Brittany	9.4	2.5	3.9	2.0	3.7	2.2

Note: a. including foremen

Source: INSEE (Structure of Employment Inquiry).

commutation, the effects of which were felt from 1977-78 on, cut manpower requirements by four. Furthermore, the national installation plan for electronic commutation came to an end at the beginning of the 1980s, thus reducing the need for exchanges and cabling and it therefore became necessary to switch to new up-and-coming sectors: for example, office automation data processing and optical fibres. The larger companies started doing just this. Thomson, for example, switched to private telephones and mini-computers. LMT, a subsidiary of Thomson in Lannion, went from the manufacture of cables to that of optical fibres. AOIP, in Morlaix, started specialising in electronic materials for professional use. It became crucial to retrain semi-skilled workers as technicians and, of the latter, electrical mechanics as computer technicians. Two methods were used: dismissals followed by rehiring after retraining, or retraining without dismissals. Indeed Brittany can boast a good training and educational system.

Table 7.7 shows us that results for 1975-81 were good: employment increased in Brittany's telephone industry whereas nationwide it decreased, with Paris region results worse than the national level. This development was to a large extent the result of a sharp rise in the number of technicians and even more so in the number of engineers. Likewise, the number of skilled workers rose steeply with a drop in the number of unskilled workers such that the differences with the Paris region narrowed appreciably concerning both the technical level of production labour and worker qualification (see Table 7.6). However, it is clear that in 1981 the effects of technological change and the prospects of a drop in public demand had not yet been felt. Since then, plans for reduction in output as well as for dismissals have abounded. An official study, presented in 1985, forecasts a drop in employment over the next two or three years with a possible recovery should the diversification scheme succeed. For those markets where international competition is strong, such as data processing and office automation, such diversification is very risky. However, the state plan for conversion from cables to optical fibres is providing a whole new captive market, after the commutation market, to the CGE group which now finds itself in a position of monopoly. It is highly probable that in the years to come, workers' jobs will be more affected than those of engineers and technicians (which again would

Table 7.7: Development of Employment in the Telephone Industry in Brittany and the Paris Region

	Variation in Total Workforce (%)	Variation in number of prod. engineers (%)	Variation in number of prod. techn. (%)	Variation in number of skilled workers (%)	Variation in number of unskilled workers (%)
Paris region	-23.0	-6.6	-11.5	-43.5	-58.4
Brittany	+8.8	+425.0	+154.0	+50.4	-32.4

Source: INSEE (Structure of Employment Inquiry)

narrow the gap in labour technical skill with the Paris region) for a great deal of research must be done to improve telephone exchanges and prepare for the next technological revolution, i.e. optical commutation, due to hit the industry within about ten years' time.

It would appear that voluntarist logic, creating research centres, 'grandes écoles' (higher university colleges) and large units of production has had the positive lasting effects even though decision centres are outside the region and industrial diversification has been poor whether it be from a sectorial point of view or concerning the growth of small local firms.

Let us illustrate this last point. Certain small local firms were set up to carry out sub-contract cabling. Most of them closed down but some found other openings while other new firms were established, particularly in another – much smaller – sector of industry in Brittany, that of electronic equipment for professional uses. In 1982 there were 26 small electronics firms in Brittany, 23 of which were less than ten years old. Many of them were involved in marine electronics on a subcontracting basis making electronic equipment for the French navy or other government departments. An industrial fabric is thus forming though it is still fragile as yet. Should it flourish, it will not be a place for unskilled labour because, as everywhere, electronics subcontracting is changing in Brittany. Principals are increasingly subcontracting completely fully-tested and guaranteed packages, the subcontractant becoming the constructor's partner, participating in product design, as well as being the middle-man between micro-electronics manufacturers and the user firms. This development has been promoted by ANVAR (a state organisation aimed at fostering innovation) which has helped a good many small sub-contractants to design new products themselves.

Robotics. The production of robots calls for expertise in different technological sectors: electronics, electrical and mechanical engineering (particularly for the construction of machine tools). In that case, do regions of early industrialisation specialising in these sectors have an advantage in the development of robotics as a result of their sectorial complementarity? In 1982 there were 27 firms manufacturing robots and manipulators (in the strict sense of the terms) spatially distributed as follows:

Table 7.8: Regional Distribution of Robotics Firms in 1982

Paris region	12
Rhône Alpes	3
Nord-Lorraine	1
Other regions of early industrialisation	2
West	6
South-West	1
Provence - Côte d'Azur	1
Limousin	1
	—
France	27

Source: Le Quement (1983)

What can we conclude from these figures?

1. Despite the clear lead that not surprisingly the Paris region holds, this industry is relatively evenly spread throughout the country, which is in sharp contrast with the findings of Camagni (1984) for Italy. There, the 18 robotics firms are all in the north, principally around Milan and Turin. Moreover, French firms are far from being concentrated around large agglomerations:

Paris agglomeration	12
Bordeaux and Grenoble areas	2
Average-sized towns	6
Small towns	7
	—
	27

2. Regions with low industrialisation have nine firms, that is one-third, which constitutes a high proportion and which would also seem to question the advantage of sectorial complementarity. However, study of the location of firms manufacturing components for the robotics industry gives the following results:

Table 7.9: Location of Firms Manufacturing Components for the Robotics Industry. 1982.

Paris region	9
Rhône-Alpes	5
Nord-Lorraine	0
Other regions of early industrialisation	3
West	3
South-West	0
Provence-Côte d'Azur	0
France	20

Source: Le Quément (1983)

This clarifies things. Robot constructors (whose location is shown in Table 7.8) are either subsidiaries of large groups (CGE, Matra, Renault) or small firms breaking into new niches on the market, and are mainly involved in design, exploiting technical know-how from various other sectors. Given the ease both with which information is disseminated and men circulate, there would appear to be few location constraints, enabling this activity to develop anywhere in the country even in the furthermost enclaves (e.g. Limousin). On the other hand, component manufacturers are much more concentrated in regions with a long industrial tradition where sectorial complementarity is important. Among these regions, Nord-Pas de Calais and Lorraine, where the electronics and mechanical engineering industries are not very important, have no manufacturers of components at all.

Sectorial reorientation from mechanical engineering to robotics can be quite clearly seen in two areas of traditional industry: Besançon in Franche-Comté and Mulhouse in Alsace (see Pottier and Touati 1984). The Besançon area has for a long time specialised in watch-making, an activity now in sharp decline. In 1965, one watch-making firm, Sormel, launched out into the manufacture of automatic conveyor belts for the watch-making industry. It was then absorbed by the Matra group and it became a pilot of the Matra's division of 'control and automation'. Thereafter, Sormel became a seed-bed unit: four former Sormel engineers setting up their own specialised machine firms between 1974 and 1980. The educational sector in Besançon has developed along the same lines. The Institute of Chronometry, founded in 1927 for the training of executives, was more or less ignored by the profession. It then became the 'Higher National Institute of Mechanics and Microtechnology', with one of the best automatics laboratories in France, working in conjuction with local firms to such an extent that local links between research and industry have now become essential.

The Mulhouse area specialises in machine tools and textile machinery, two sectors currently under severe strain. Here, the firm leading the field in the conversion to robotics is a large mechanical engineering firm, Manurhin, who first specialised in articulated automation and optical controls before being absorbed into the Matra group which has a good experience of micro-electronics. Mulhouse has become a centre of the automation industry with strong links between industry, research and education. Within the university of Haute Alsace, the 'Institute of Multitechnical Research' is developing process automation and there are teaching establishments where the training of electronics specialists is carried out in conjunction with the traditional training of mechanics.

Thus, for these two robotics leaders, Sormel and Manurhin, sectorial reorientation is obvious. Their absorption into the Matra group is symptomatic of the way robotics is developing, foreign competition forcing firms to manufacture complete production systems rather than individual robots. Such a necessity could force smaller firms to merge into larger groups who alone can pursue a strategy concerning the product process automation sector as a whole. Such merging could entice robot constructors into industrial areas where complementary firms may be found. However, nothing is preventing small innovative firms from continuing their development throughout the country, even if they are later absorbed by large groups.

Aeronautics. Aeronautics is a high technology industry where France is a world leader. It is a downside sector involving many new technological activities such as electronic components, computer-aided design and manufacture as well as new materials. Research accounts for 20 per cent of turnover, a figure which is much higher than in other industrial sectors. In 1981, engineers and technicians accounted for 45 per cent of all production personnel and 90 per cent of the workers were skilled. The aeronautics industry is made up of some 150 firms, 25 working in air-frames and engines (mostly large state-owned firms) and 125 small and average-sized firms working in fittings. New technology is also important for these latter companies. This is obviously the case for equipment suppliers specialising in electronics but also for those who, especially over the last five years, have had to replace special steel with composite materials.

Let us consider Clausse's analysis (1984) of location of the French aeronautics industry. The present siting is for the most part the result of the strategic decentralisation policy pursued between the wars and after in the 1960s. A part of large firm activity was transferred away from the Paris region to other large agglomerations, notably in the South (Toulouse, Bordeaux and Marseille) as well as to Nantes - Saint Nazaire. However, the urban hierarchy was respected in this decentralisation process:

1. Decision-making, research and training remained concentrated in the Paris region.
2. The training of engineers, technicians and skilled workers in state-run establishments was decentralised only to Toulouse.
3. Research developed in Toulouse, Bordeaux and Marseille.
4. However, testing and production activities were not limited to these large towns. They were also located in the Nantes - Saint Nazaire area and in average-sized towns in the South-West: Tarbes, Pau and Diarritz.
5. Within these production activities, several different kinds of job need to be distinguished:
   - final assembly is carried out in the large and average-sized towns just mentioned;
   - intermediate assembly and initial production are carried out near to final assembly except when such activities are part of an international co-operation agreement;
   - finally, subcontractants and equipment suppliers are to be found throughout the country, including rural areas, with two poles of concentration in the Adour valley near Biarritz and at the foot of the Pyrenees.

Along with this urban hierarchy, other criteria are to be taken into consideration for the spatial distribution of the sector: first of all, spatial specialisation based on the nature of the product involved, due to the important role played by final assembly in the polarisation of activities. Thus Toulouse specialises in passenger airliners, Bordeaux in fighter planes and Marseille in helicopters. Moreover, a veritable regional production network has been organised in the South-West as seen for subcontractants and equipment suppliers. However, this regional network is not shut off from other regions and especially not from abroad in the context of growing co-operation with European countries. For example, the Dassault group, in its production plant in Biarritz, works in close conjunction with Spanish firms.

Table 7.10 shows that the technical skill in production jobs is much higher in the Paris region than elsewhere, except in Provence - Côte d'Azur (Marseille area), while it is about twice as low in 'other regions of early industrialisation' and in the West. The South-West is for its part half-way between the two. In 1981, the gap between the Paris region and the national average narrowed only slightly (the ratio went from 1.24 to 1.23) but the level of worker qualification in the Paris region approached that of the South. The decision-making predominance of the Paris region is testified by the number of employees not involved in production: 27 per cent of all employees in 1975 compared with 20 per cent for the South. Two factors would appear to be responsible for the Paris region's continuing predominance:

1. The development of computerised communications between establishments, allowing the region to maintain its position as leader. For example, Dassault's cam/cad systems for its production plants are linked up to data banks in the Paris region.
2. The size of the Parisian network of specialised-equipment suppliers in high technology electronics, principally the result of a large market and ready availability of top engineers and technicians.

Conclusion
The following conclusions may be drawn from the two approaches, global and sectorial, we have studied:

1. High technology industry is spreading throughout France, corresponding to the growing need for innovation in firms as a whole in a

Table 7.10: Total Workforce, Degree of Technical Skill and Worker Qualification in the Aeronautics and Space Industry

	Total Workforce		Technical Skill (a) in Product. Work		Worker Qualification (b)	
	1981	Variation 1975-1982 (%)	1975	1981	1975	1981
Paris region	49,000	+13	51.3	55.0	86.5	91.0
Rhone-Alpes	4,700	+195	20.1	19.9	56.3	79.1
Nord-Lorraine	500	-	-	-	-	-
Other reg. of early industrialisation	6,000	+65	26.2	29.3	84.3	90.1
West	14,000	0	25.9	28.4	85.6	83.3
South-West	28,000	0	37.3	40.9	92.5	94.6
Provence-C.d'Azur	9,400	+16	53.4	51.7	91.9	93.6
France	112,000	-13	41.4	44.6	86.8	89.7

Notes: a. Number of engineers and technicians as a % of production employment.
b. Number of skilled workers as a % of total number of manual workers.

Sources: Ministry of Industry: Annual Inquiry into Industrial Firms (total workforce). INSEE: Structure of Employment Inquiry (technical skill and worker qualification).

context of technical and market instability. This can be seen in the number of professional and state-run technical aid centres now available to firms. But, in relative magnitude, this process is only of advantage to Western regions and Provence - Côte d'Azur. In 1975 in the West, technical skill in production labour was the lowest and R&D among the lowest, but between 1975 and 1982 the latter developed much more than in other regions. Thus the number of engineers and technicians has risen sharply in the Brittany telephone industry, while robotics firms are to be found in the western regions. During the 1960s decentralisation process, these regions had specialised in assembly but with the transfer of part of this sector to the third world, the degree of labour technical skill has risen.

2.   With balancing out limited to only certain regions, Paris's domination is in no way under challenge. Indeed, the sharp drop in the number of manual workers in this region has led to a strengthening of its specialisation in industrial R&D. Aeronautics is an example of how telematics can contribute to centralisation. Similarly there is a constant rapport between the degree of technical skill and the size of urban unit, with even a widening of the gulf at 50,000 inhabitants. This goes along with the conclusions drawn by Oakey, Thwaites and Nash (1980) and Malecki (1983) concerning the permanence of R&D concentration in Great Britain and the United States.

3.   A pleasant environment is an important criterion in the location of high technology industry as seen for Provence - Côte d'Azur where the numbers of R&D personnel, which was already high in 1975, increased faster than the national average from 1975 to 1982, the result of limited constraints put on high technology location when seeking to satisfy engineers and managers' desire for a better quality of life. Natural and cultural aspects to the attraction of the environment must therefore be taken into account as indeed must convenient travel facilities, especially if the activity is involved in the international division of labour as we have seen with micro-electronics. In Provence-Côte d'Azur, the size and growth of industrial R&D is concomitant with high immigration and the large number of new firms.

However, these two facts would not seem to be correlated after calculations made for the 21 regions.

4. On the other hand, areas of early industrialisation suffer from a much more unpleasant environment. Here it seems that the whole social set-up is obsolete but we must not be too hasty in condemning its ability to develop. It is obvious that sectorial specialisation has an important role to play. For example, R&D growth is much slower in Nord-Pas de Calais and Lorraine than in other regions of early industrialisation. Of the latter, it has been seen that those specialising in mechanical engineering are now turning to robotics.

References

Aydalot, Ph. (1985) 'L'aptitude des milieux locaux à promouvoir l'innovation technologique'. Paper presented at the meeting on 'New Technologies', SRBII-ASRDLF, Bruxelles

Camagni, R. (ed.) (1984) Il robot italiano, Edizioni del Sole - 24 Ore, Milano

Clausse, J.R. (1984) 'Rapport sur le fonctionnement de l'industrie aéronautique française'. Université de Paris I

Delpierre, M. (1985) 'Les mutations spatiales dans la crise ...' Revue Française des Affaires Sociales, no. 1

Le Quément, J. (1983) L'Usine du futur proche, Agence de l'Informatique. Hermès, Paris

Malecki, E.J. (1983) 'Technology and Regional Economic Change: Trends and Prospects for Developed Economies.' Paper presented at the 'North American Meeting of the Regional Science Association', Chicago

Nishioka, H. (1983) 'High Technology Industry and Regional Development'. Case study for the workshop on 'Research, Technology and Regional Policy', OECD

Oakey, R.P., Thwaites, A.T. and Nash, P.R. (1980) 'The Regional Distribution of Innovative Manufacturing Establishments in Britain', Regional Studies, Vol. 14

Planque, B. (1983) Innovation et développement régional. Economica, Paris

Pottier, C. (1983) 'Les conditions d'utilisation de la main-d'oeuvre comme facteurs de localisation des groupes industriels'. Mondes en Développement, no. 43/44.

Pottier, C. and Touati, P.Y. (1981) 'Concurrence internationale et localisation de l'industrie des semi-conducteurs en France'. Dossiers du C3E. no. 24, Université de Paris I

Pottier, C. and Touati, P.Y. (1984) 'L'évolution des industries régionales en France face aux mutations technologiques'. Dossiers du C3E, no. 43, Université de Paris I

Saint-Julien, T. (1984) 'La crise et l'évolution de la qualification du travail dans les villes françaises'. Université de Paris I. See also in <u>Revue d'Economie Régionale et Urbaine</u>, 1985, no. 1

Chapter Eight

HIGH TECHNOLOGY INDUSTRY IN AUSTRALIA: A MATTER OF POLICY

Stuart Macdonald

Introduction

Like beauty, Australian high technology is largely in the mind of the beholder. There is as little as would justify policy prescriptions to encourage high technology, and as much as would prove the success of that policy. Australia lacks any consensus in the definition of high technology, and probably the statistics to measure the activity accurately even were there a consensus. There is a profusion of high technology policy in Australia, all aimed at stimulating whatever looks like high technology. Without policy, it is claimed, there will be no high technology industry in Australia, and in a land where there seems to be policy for nearly all industrial activity, perhaps this is so. Certainly it is assumed that where the policy is right there will be a high technology industry, and sheer amount of policy is even seen as a suitable indicator of the existence of the industry. That is not to say that all high technology policy in Australia is effective - in fact, much of it is decidedly inappropriate - but policy does offer an approach to a problem that is otherwise intractible. There are no comprehensive studies of high technology industry in Australia and even case studies are rare. Consequently, it is from a policy perspective that the development of high technology industry in Australia must be viewed.

Defining High Technology

Part of the reason why there is no commonly accepted definition of high technology is because everybody knows instinctively what it is. Because instincts vary, so do definitions. The Australian Department of Science and Technology - which has also been known to rely upon its instincts - considers high

technology industries to be those which indulge in
generous research and development (R&D) spending, in
which there is interaction between scientific and
technical skills, which generate new products and
processes, and in which there are entrepreneurs with
scientific or technological backgrounds (Department
of Science of Technology, 1981, page 3). Such
indicators, like most of those used to distinguish
high technology from other industrial activity,
catch the flavour of high technology by concentrating
on major ingredients. The recipe, however, remains
elusive (see Carlson, 1984). When reduced from theory
to the practice of a standard industrial
classification, such ingredients yield decidedly
unsatisfying fare. For instance, they virtually
ignore the whole of the tertiary sector because its
investigative activity is scarcely acknowledged to
be R&D in Australia (see Macdonald, 1983a). Moreover,
even within the manufacturing sector such an approach
can produce totally conflicting results. In the
United States, where there is now no shortage of
lists of high technology industries, one major study
includes rubber and oil among high technology
industries (Armington, Harris and Olde, 1983):
another categorically declares, 'High technology
entrepreneurs are a breed apart from the
entrepreneurs of traditional industries like steel,
rubber, oil and automobiles' (Premus, 1983, page
56). While such industry lists fail utterly to be
definitive, and have not apparently been used in
Australia anyway, they do at least discourage more
fanciful definitions. Among the several of these with
some currency in Australia at the moment is that from
the Chairman of the Commonwealth Scientific and
Industrial Research Organisation (CSIRO), the
country's largest research group:

> ... my definition of high technology is that
> which originates from science and scientific
> research - research derived through what we call
> the scientific method. (Wild, 1984, pages 47-8)

Classifying industries by the purity of their
research would pose some fascinating problems.
     There is an alternative approach, though still
not one which yields a definitive taxonomy of high
technology industries. High technology is high not
because it is nearer to God than ordinary technology,
but because it involves high risk and possibly high
return, a high rate of change, and - especially -
high information intensity. That definition is

obviously inadequate - it could apply quite nicely to many sophisticated criminal activities - but it does have the merit of stressing the utter dependence of high technology industry on its ability to handle vast quantities of new information. Such a notion is quite foreign to most definitions of high technology, except in as much as they are preoccupied with what may be a very restricted information activity - research. Neither is this fundamental characteristic considered even relevant to much of the high technology policy currently so prevalent in Australia. Yet, their distinctive use of information is the common factor among high technology firms and it is what makes them different from the mass of firms engaged in most other industrial activity. Certainly they have little else in common and to speak of high technology industry as if it were a single entity is not always helpful. For this reason there may, in fact, be little profit in seeking a single, comprehensive definition of high technology industry. The industry comprises tiny firms and huge transnationals exploiting different technologies for different purposes, and often in conjunction with established technologies; it can emerge in any part of the spectrum of existing industrial activity, and the characteristics of its individual firms may change rapidly. In practice, high technology firms have few characteristics in common except high information intensity, and that is difficult to measure rigorously and is not exclusive to high technology anyway.

Benefits and Costs of High Technology

High technology - whatever it is - is a rare and precious economic activity in that it is widely assumed to deliver prodigious benefits while imposing almost no costs at all. With a benefit:cost ratio approaching infinity, high technology has captivated those responsible for the economic management of whole nations, and whole regions, and whole towns, and whole villages. Such unbounded political enthusiasm is perhaps surprising: political fingers were just about all that was even singed by the 'white heat of the technological revolution', and it seems only yesterday that technology was firmly cast in the role of despoiler of the workforce. But the easy benefits claimed for high technology have proved irresistible, as have the apparently low costs, not just of high technology industry in action, but of programmes to stimulate

that action. In fact, the benefits to be gained from
high technology have probably been grossly over-
estimated, and the costs - especially the indirect
and hidden costs - may actually be quite
considerable. Such costs may accompany the benefits
that are realised, or may be the corollary of
benefits which do not eventuate. In Australia, they
may often be the indirect costs of high technology
policy itself.

Employment

Most prominent by far among the benefits claimed for
high technology is its apparent ability to create
employment in the depths of prolonged recession. When
the responsible Australian Minister was frustrated
in his plans to entice an American semiconductor firm
to locate in Canberra recently, he declared that this
single plant would have led

> to the establishment ..., without any shadow of
> a doubt, of an applications industry with a
> capacity to employ in excess of 100,000 people
> at the end of the decade. (Hodgman, 1981, page
> 951)

That would have been almost 2 per cent of the entire
Australian workforce. In reality, high technology
industry is not a large employer now and there is no
prospect of it becoming one in any future that is
even dimly foreseeable (Rumberger and Levin, 1984;
Riche, Hecker and Burgan, 1983). Real job growth has
been in the tertiary sector generally, and especially
in some of the less salubrious service functions.
U.S. employment in the fast food industry is greater
than in high technology, and it has grown faster both
proportionally and absolutely - all without a hint of
policy to encourage the consumption of hamburgers
(Science and Government Report, 1983; Windschuttle,
1984). Moreover, it is in such unglamorous sectors
that most future employment growth is forecast. In
the Australian State of Queensland, the agriculture,
mining and manufacturing sectors combined have been
responsible for only 5 per cent of new jobs over the
last ten years - despite government policy dedicated
to creating employment in those very sectors
(Mandeville, Macdonald, Thompson and Lamberton,
1983).
    Of course, it could be argued - and is argued -
that low-grade service functions create jobs mainly
for the semi-skilled and unskilled, and do not make

full use of human capital. Unfortunately, the same is true of much high technology employment. That is why parts of the American industry are located in cheap-labour countries. Highly-qualified professional employees may be more common in high technology than in other industries, but the typical employee in the United States, is poorly-paid, part-time and female, with the daily prospect of tedious, repetitive work, with no union protection and no career structure (Tamaskovic-Devey and Miller, 1983). There is even substantial and illicit 'back-alley' employment associated with some high technology industry.

The products of high technology industry may, of course, help generate employment in other industries altogether. They may equally well reduce that employment by allowing what has become known as 'jobless growth'. It all depends on how the technology is used and on the facility with which organisations adapt to change (Mandeville and Macdonald, 1983). If firms in established industries cannot seize opportunities presented by new technology, or if they adopt that technology with no other intention than to cut labour costs, then high technology may very well reduce rather than increase overall employment. Certainly there is no justification for the bold assumption that high technology necessarily generates employment in the rest of the economy (see Markusen, 1983).

Wealth generation and growth

A perpetual dilemma is faced by those whose political life cannot be guaranteed to extend beyond the time taken for policies to yield their benefits. The problem is thought unlikely to arise with high technology because returns are imagined to be not only huge, but immediate. Certainly the industry has grown, and continues to grow, rapidly, though from a small base. While it is incontrovertible that some high technology firms have flourished beyond perhaps even the imaginings of their founders, and have made fortunes for their owners, that is not typical. Many high technology firms fail, most just linger, and only a very few have instant success. The few, however, are often presented in Australia as if their fate were inevitable, as if a major share of the international market is always assured, as if the generation of spin-off firms somehow defies the commercial laws of natural selection. That is most misleading. Moreover, the wealth that is undoubtedly created in some quantity by these few, is likely to be poorly distributed. Whatever the benefits of wealth generation, they should be discounted for the social costs of their mal-distribution.

Environmental and health advantages
Unlike the 'smokestack' industries with which they are commonly compared, high technology industries are supposed to be clean and healthy, a boon to their employees and the environment. Just why comparison should always be made with the older parts of manufacturing industry, which now provide relatively little employment, is something of a mystery. Comparison with other growth sectors, with office or even fast-food employment, might be more meaningful, though less impressive. But the smokestack industries offer ample visible evidence of pollution and poor working conditions: in high technology the evidence is simply less visible. In fact, chemical pollution is now a serious problem in the Mecca of high technology, Silicon Valley (LaDou, 1984; Begley and Carey, 1984; Soiffer, 1984). So, too, are all those problems traditionally associated with smokestack industries - over-crowding, traffic congestion, smog, poverty and even hunger (Kutzman, 1984; Dianda, 1984; Rudy, 1984). It is quite wrong to imagine that thriving high technology is kind to the environment and gentle with humanity (Rogers and Larsen, 1984).

Restructuring of the economy
Of all the proclaimed benefits of high technology, its role in the restructuring of developed economies is certainly the most important and probably the least contentious. The primary and secondary sectors have long ceased to produce enough wealth, or the right sort of wealth, to satisfy the requirements of modern society. The tertiary sector - the service sector - has grown prodigiously and now dwarfs the other two. Its function is not only to organise the production of wealth, but also to produce its own, though in an intangible form not always familiar as wealth. Much of this wealth, and a good deal of the wealth produced within the secondary sector and even the primary sector, is in the form of information.
     The diffusion of information technology is only an indicator of the information revolution that is taking place; the force of the revolution is information itself. Information workers comprise between a third and a half of the workforce in

developed economies (OECD, 1981; Porat, 1978, page 4). But in proportion to its importance as the major form of wealth in a developed economy, and a prerequisite for the production of all other wealth, far too little is known about information and its organisation. Government policy, in particular, seems to be crippled by an attitude which still regards the manufacturing sector as the only possible engine of growth, which recognises wealth only in tangible form, and which values information as just another input to the production of 'real' wealth (see Mandeville, Macdonald, Thompson and Lamberton, 1983). Government high technology policy in Australia is no exception. High technology industry uses information resources that many other industries have either neglected or used inefficiently (see Lamberton, Macdonald and Mandeville, 1982). Other parts of the economy - perhaps all other parts - must acquire the same facility, and must undergo the reassessment and re-organisation resulting from the new utilisation of a major resource. High technology industry, by example, by its own growth, and by the diffusion of its products through an economy, can play an essential leadership role in what is proving to be a difficult and painful restructuring process.

Neglect of comparative advantage
Because the market is thought unlikely to work well in such a new and uncertain area as high technology, governments feel compelled to interfere. Governments influence decisions on where limited resources may best be applied, and they should also consider just where the comparative advantage of their constituencies lies. Advantages that are claimed by State governments and local authorities in Australia are often impossible to take seriously. Clean water is said to give Canberra an edge, in Tasmania it is pure air, in Perth and Adelaide tertiary education, in Newcastle and Wollongong the existing electronics industry, and in Queensland 'bright people' and an 'established reputation' (see Maconald, 1983b). Governments have simply assumed that their constituencies have an obvious comparative advantage in high technology and - attracted by its assumed benefits - are anxious to direct resources towards that activity (see Gregerman, 1984).

Too often the information requirements of high technology industry are either ignored in policy, or are not afforded the central position they deserve.

Too often those responsible for policy have only a vague idea of what high technology is. High technology industry is seen to be whatever fills Silicon Valley and so the policy task becomes the replication of Silicon Valleys in all sorts of unlikely spots. What is missed in all this energetic innocence is that the most outstanding concentrations of high technology industry were planned by no one (see Dorfman, 1983), though there are those who would apply policy retrospectively and attribute Silicon Valley to the ambitions of Frederick Terman and the presence of Stanford Industrial Park (e.g., Weber, 1981; Beyers, 1983). In fact, the Industrial Park is a product of the Valley's development and not vice versa (Board of Trustees, 1974). It is, of course, possible to mimic the obvious characteristics of Silicon Valley elsewhere, to construct what are really no more than Hollywood film sets of Silicon Valley. Those characteristics can hardly have been missed by the dozens of policy makers from Australia who have been conducted through the Valley, though its less desirable attributes - the expensive housing, the uncertain electricity and water supplies, the chemical pollution, the smog, and a variety of serious social problems - do seem to have been overlooked (see Rogers and Larsen, 1984). What are also missed, because they are intangible and no one is really looking for them anyway, are the intricate networks of surging information channels which supply high technology industry with its basic requirement. Key personnel in high technology know other industry experts and mobility is high. For such people, work substitutes for religion and success for heaven. That is attained through a vicious dedication to acquiring and using information. In Australia, such matters are not considered to contribute to comparative advantage in high technology and are not thought relevant to high technology policy.

It is axiomatic that State government claims of comparative advantage for high technology industry cannot all be correct, and to the extent that they are wrong, costs are likely to be incurred. Indeed, it is unlikely that Australia offers any overall comparative advantage to many high technology firms, and incentives should be seen for what they often are - government compensation for lack of comparative advantage. Beyond the direct - and probably fairly moderate - cost of such incentives, there are likely to be further and greater costs associated both with high technology industries encouraged to locate in the wrong place, and with the neglect of less glamorous industries. Without careful assessment of high technology policies, the precise magnitude of these costs is likely to remain hidden. The euphoria for high technology policy in Australia - with politicians and bureaucrats proclaiming comparative advantage in biotechnology, microelectronics and information technology for just about anywhere - is not conducive to such careful assessment.

Inappropriate policy

If high technology is important, then it is also important that high technology policy be correct - rather than simply highly visible - if opportunities are not to be missed and resources wasted (see Ballinger, Hope and Utterback, 1983). There are costs associated with ill-conceived policy, and probably particularly with the general assumption that high technology is a single industry with uniform requirements. It is not, and firms are likely to require programmes tailored to their particular circumstances. They also require co-ordination of high technology policy with other policies (see Office of Technology of Assessment, 1984). Federal policy to encourage high technology firms may be frustrated by State policy to attract those firms, however inappropriate the location, or by other Federal policy persuading firms to go to areas of, say, high unemployment (see Economist, 1983). Similarly, there are obvious costs when one region or one city competes for high technology industry by out-bidding the incentives of others. That is parish-pump mercantilism, and literally a policy for high technology at any cost. But there may be competing policies even within a single government as individual departments seek to exploit their own high technology policies to further their cause in internecine power struggles. In Australia, where State governments offer incentives to discourage high technology firms from locating in other States and thereby negate much Federal high technology policy, competition certainly descends to departmental level (see Feller, 1984).

Market distortion

There is but a small step from regarding high technology as a wealth-creating activity increasing the prosperity of the economy, to regarding high technology as a cost to be borne for the prosperity of the economy. Government incentives for high technology are subsidies and - much like tariff protection - help to shelter that industry from competition. The spectre of highly-protected high technology looms large in Australia and that sort of industry would undoubtedly impose severe costs on the rest of the economy. Not only would high technology purchases be more expensive, but high technology's leadership role in restructuring would be substantially weakened, and a new distortion in the market would deprive alternative activities of resources they would otherwise have attracted. To the extent that policy distorts what is already a highly imperfect high technology market, it makes more confusing the market signals on which business decisions must be made. Similarly, the rampant speculation in high technology shares now occurring in Australia is not altogether a cause for celebration; it confuses the market and makes investment hazardous. As one leading stockbroker has observed, it is reminiscent of the speculation in 'penny dreadful' oil and mineral shares in the early seventies (Ord Minnet, 1985). That is not a sound basis for the restructuring of the Australian economy.

Information and High Technology Industry
The innovative process - in every industry - is largely an information process (see Macdonald, 1983a). Information is gathered from a multitude of sources and occasionally new information is created - most obviously during research, but also during other phases of the innovative process, such as development and marketing. The assembly of various bits of information into a new pattern is the essence of a new product (or process), whether that product takes tangible or intangible form. In many mature industries, the marginal information content of a new product over its predecessor may be small, and the information pattern may be little changed. Large firms within such industries may already possess much of the information required for this new pattern. They may find, however, that their own complex and rigid organisational forms tend to inhibit the assembly of new information patterns. Indeed, large firms, especially in established industries, are often castigated for their lack of innovative zeal, although this is usually attributed to low incentive to innovate rather than to their inability to exploit information resources.
    High technology industry is markedly different,

as might be surmised from the importance of small, new firms in that industry. It is not that a high technology firm necessarily requires more information than other firms in order to innovate, but it does require more information relative to other inputs. Its activities are information intensive; so much so, in fact, that it cannot possibly rely on internal sources for all the information required. The high technology firm must actively seek information outside, and is dependent on the efficiency of its information channels to the world beyond. The characteristics of information make it a difficult subject of negotiation and hard to price, so that information is commonly exchanged for other information and channels must conduct information in both directions. The high technology firm must supply information in order to procure information, though what it supplies need not be from its own internal sources.

There are formal channels through which information travels. Patent specifications, licences, books, journal articles, trade publications and conference papers are obvious examples. Information may also be embodied in the hardware a firm produces and sells. None of these, however, guarantees that information transferred will be strictly relevant and comprehensive - or sometimes even comprehensible. There are other, informal channels through which information can flow which are able to compensate for the inadequacies of formal channels. The importance of personal contact as a means of gaining and supplying full and specific information has long been recognised in studies of the process of technological change (e.g. Teece, 1976; Cetron and Davidson, 1977; Macdonald, 1983c). It is not argued that personal, informal contact is a substitute for more formal means of information transfer, but that it is an essential complement.

High technology industry in the United States is heavily reliant on personal, informal channels for information flow. Key employees have their own networks of contacts who readily supply information which is not obviously proprietary in exchange for other information. Personnel mobility has long been high in the industry as employees take advantage of the information they possess and carry it with them to new jobs (see Baram, 1968). Very often the information which flows in this way is commercial rather than strictly technical because it is essential to the success of high technology industry that information from the market be mixed as early and as thoroughly as possible with technical information. There is now even a substantial high technology 'headhunting' industry in Silicon Valley, a result of the growing demand for key employees at a time when high technology industry has grown too large to be totally embraced by a single, informal information network. Headhunters are most frequently commissioned to search for individuals who possess both technical awareness and related commercial expertise, the latter commodity being rather more rare and precious than the former. Very often the information that is most valued by high technology industry is apparently insignificant, more related to practical experience and inspiration than to the sort of information acquired for a university degree.

The way in which information travels in high technology industry in the United States is partly cause and partly consequence of that industry's frantic pace of technological change. Because information can flow fully and rapidly when there are personal and informal channels, change is swift. Because change is swift, there is a need for information to flow fully and rapidly. The one begets the other. U.S. high technology industry continually adjusts to change; even the semiconductor industry - now thirty years old - has refused to display the characteristics of a stolid, mature industry (Braun and Macdonald, 1982). Most high technology industry in most other countries - especially the large firms which qualify for so much policy support in those countries - has great trouble keeping pace (see Economist, 1984). The problem is, in part, an infrastructure which is not conducive to rapid information flow along personal and informal channels. Only the Japanese, forced to compensate for the total absence of such channels with a completely different system of organised information flow, have overcome this obstacle. In Australia and many other countries, and even in those parts of the United States distant from centres of high technology, there is inadequate recognition of the importance of information to high technology, of the personal and informal channels through which much information must flow, and of an infrastructure conducive to the flow of information in this manner.

Infrastructure for High Technology in Australia
Australia's reputation overseas as a land of sturdy individualism, a last frontier where the entrepreneur may thrive, a place sufficiently

developed to exploit massive undeveloped resources, is largely undeserved and thoroughly misleading. Odd attitudes are nurtured and harboured here, strange situations have developed through isolation and have become familiar with time, peculiar policy has been tolerated long enough to become acceptable. This distinctive culture is not entirely conducive to the development of high technology industry.

Perhaps the main impediment to high technology growth is the distortion in the rest of the Australian economy created by extensive government involvement. Traditional industries in the manufacturing sector are heavily protected, and the attitude that it is the responsibility of the government to ensure industrial prosperity is widespread. Even high technology industry is not immune from the contagion, and there are new industry associations which lobby for government assistance. They give examples of government support of high technology overseas, they stress the importance of high technology for Australia's defence, the need for a high technology infrastructure for the rest of the economy, and they even resort to traditional infant industry arguments (e.g., Australian Computer Equipment Manufacturers' Association, 1983; Durie, 1981). Much less evident is the notion that high technology industry in Australia should be based on some comparative advantage and should aim to be internationally competitive. Consider the opinion of one leader of the Australian electronics industry, an industry which is commonly regarded as a high technology stalwart:

> ... I cannot feel ashamed to suggest that an industry of the nature of the electronics industry deserves protection, and then not temporarily but permanently. (Huyer, 1974)

That sort of attitude has helped create an interesting breed of Australian entrepreneurs, one which - quite naturally - is as dedicated to exploiting the opportunities presented by government as those presented by the market. As one entrepreneur in the computer industry explained:

> I find it irrational to refuse these debilitating handouts, which are so easily obtained. As a salve to the conscience I use the fact that my taxes contribute to this endless pool of money, and that refusal would be a disadvantage against the competition which

accepts. (Webster, 1980)

As for the Australian electronics industry, only a decade ago that industry was still making semiconductor components using 1950s technology and, despite Australia's tiny market, employed 1,000 per cent effective tariff protection to manufacture a greater range of components than nearly any other country (Rattigan, 1977).

The attitude that the government should be responsible for whatever has to be done extends even to research and development (R&D). In Australia, the government not only pays for nearly 80 per cent of all R&D, it actually performs that proportion. The measured R&D performed by the private sector is tiny and has actually been contracting in recent years despite, or perhaps because of, increased government incentives (see Gannicott, 1980a). CSIRO alone now performs more R&D than all industries in the manufacturing sector combined (Macdonald, 1982–83). Despite frequent exhortation to undertake more research, the private sector remains stolidly unenthusiastic. Consider this statement from a major Australian industries association:

> Industrial R&D funds come almost without exception from a company's gross profit. It is a risky business. To commit shareholders (sic) funds or loan moneys to this kind of venture brings in the risk of damaging the enterprise as a whole. This would be an improbable step for responsible and efficient management. (Australian Industries Development Association, 1974, page 7)

Supporting, perhaps explaining, this unusual Australian research structure is a widespread belief in the validity of the linear model of science policy. The innovative process is imagined to start with invention and to proceed in an orderly sort of way to innovation, basic research is followed by applied research and then by development and marketing, science produces technology. Such a model may be convenient, but it is not realistic. It has been observed, for example, that much invention in fact springs from innovation, that basic research may be a response to problems encountered in applied research, that the market may precipitate research, that much science leads only to more science and that technology tends to emerge, not from science, but from other technology (Gibbons and Johnston, 1974;

Encel, 1970; Gannicott, 1980b).

The main consequence of tacit, and sometimes abject, adherence to the linear model is over-concentration on the earlier parts of the innovative process in the belief that once the ball is set rolling its own momentum will carry it some distance. Much less interest is shown in the later stages of the innovative process (see Gannicott, 1980a; Green and Ronayne, 1980; Gannicott 1980c; Green and Ronayne, 1981). This pattern is intensified by an academic disdain for what are taken to be the least intellectual parts of the process, a disdain which does nothing to improve the flow of information between universities and industry.

Unfortunately, the current fervour for high technology in Australia has re-kindled faith in the linear model. High technology is imagined to be a distinctly scientific activity and therefore dependent on basic research and universities. In fact, high technology industry is non-scientific, even anti-science in its desperation to avoid problems and undirected enquiry; it seeks the quickest and easiest route possible through what is so expressively termed 'black box' technology. The aim is to make the box work for someone without ever having to open it. Certainly high technology industry requires scientific information, but it shows little interest in creating its own, and its demands for other sorts of information are very much greater.

The dynamic nature of high technology industry demands mobility of resources. Australia lacks that mobility, especially in the most essential high technology resource - manpower. Over three-quarters of the nation's R&D manpower is accustomed to high status, superannuated job security and to gentlemanly, slowly evolving research programmes. This is not the obvious raw material for risky, cut-throat and volatile high technology industry. There is little evidence in Australia of the personnel mobility - of headhunting, for example - vital in high technology industry for the transfer of information 'on the hoof' (Senate Standing Committee on Science and the Environment, 1979). Consequently, information remains isolated: it is quite clear from extensive surveys of the private sector that no part of government or university research is regarded as an important source of technological information (Macdonald and Mandevelle, 1982). There could be no greater indictment of that research, no matter what its 'quality'. Efforts to prise open information channels through employee secondment or contracting out have been no more than platitudinous. There is no concentration of high technology industry in Australia to make mobility easier, it is difficult to transfer superannuation entitlements (see Australian Science and Technology Council, 1978), and employee stock options - which are essential elsewhere in attracting talent to new firms - are hardly thought relevant to high technology policy (see Macdonald, 1979).

The complacency of the Australian private sector was well matched by the almost inert interest in high technology shown by the conservative Fraser government between 1975 and 1983. Those years saw a succession of Industry Ministers dedicated mainly to preserving the established structure, and of Science and Technology Ministers most notable for their ignorance of matters scientific or technological. Despite every indication to the contrary, the last of these declared,

> To establish new technology-based companies a country must have a well educated population, a sound industrial base, a stable economy and an adequate pool of capital. Within Australia we have all of these qualities ... (Thomson, 1983, page 4)

That same Minister had been known to confuse his technologies by spelling silicon with an 'e' (e.g. Commonwealth Record, 1982), which is no more shameful than the admission by the Ministry for Industry and Commerce in 1981 that he did not know what biotechnology was (Lynch, 1981). Not surprisingly, the importance of information to high technology required a conceptual leap entirely beyond this government and Australia declined to participate in the OECD survey of national information economies (Lamberton, 1984). Only the threat of defeat in the elections of March 1983 prompted the Fraser government to take a sudden, belated, interest in high technology - in January (Fraser, 1983; Heywood, 1983). By that time the Labor opposition had exploited for more than a year the considerable electoral capital in high technology as a panacea for all manner of economic maladies. To Australians, discontented with a government which persisted in its conviction that high unemployment was a remedy for high inflation, high technology was an attractive alternative.

High Tech Hype

The Labor incumbent to the Science and Technology
portfolio was Barry Jones, a knowledgeable and
passionate advocate of high technology. The new
Minister seized every opportunity to bring to public
attention the need for high technology industry in
Australia, the importance of recognising the
transition to an information economy, and the
fundamental inability of the existing national
infrastructure to support high technology industry
(see Jones, 1982). When the Prime Minister's national
economic summit meeting (designed to achieve
consensus among the chief participants in the
economy) failed to emphasise the importance of
technology, Jones arranged his own national
technology summit to do just that (see Department of
Science and Technology, 1984a). It was to be only one
of many events staged to assert the importance of
high technology.

Initially, the general economic strategy of the
Labor government was well able to accommodate the
enthusiasm with which Jones threw himself into his
task. The stagnant economic policy of the previous
conservative government had offended the electorate,
and radical alternatives were welcomed. But change
always threatens established interests and is
particularly threatening when it involves the re-
allocation of resources. Neither existing manufacturing industry, heavily dependent on assistance from
Canberra, nor the scientific establishment, reliant
on government for nearly all its funding, shared the
Minister's conviction that urgent and drastic action
was required. Pressure mounted for some tempering of
the Minister's immoderate and impolitic stance. In
late 1984, the Technology portfolio was taken from
Jones and given to the Minister for Industry and
Commerce, a clever shuffling which put both Jones and
Technology in their places. High technology matters
are now the concern of a Department which has
previously shown only the slightest interest in any
sort of technology, and high technology's champion
retains responsibility for only Science, an area
which has been denied the patronage to which it feels
traditionally entitled.

Jones' contribution to Australian high
technology policy has been fundamental - literally in
the sense that he attacked the existing scientific
and industrial structure of the country. He went so
far as to identify 16 specific 'sunrise' industries
and confessed that government targeting of resources
was both necessary and desirable. That was not

popular among those unlikely to be targeted, nor
probably with those public servants left wriggling
with the argument that in choosing sunrise industries
the government was merely finding 'key industries' to
fortify Australia's technological infrastructure,
and was not actually picking winners (Tegart, 1983).
In order to use resources more efficiently, Jones
directed that portions of some existing budgets be
diverted to high technology. That was not popular
among those with existing programmes and accustomed
to greater independence in resource allocation. He
even went so far as to criticise the effectiveness of
CSIRO research in a bold attempt to revitalise that
sacrosanct institution. He recognised some of the
main characteristics of high technology industry,
particularly its information intensity, and was
consequently forced to acknowledge that the existing
Australian infrastructure was quite unsuited to such
an activity. The high public profile he gave high
technology was his attempt to gain acceptance that
the basic infrastructure required change. That, of
course, was not allowed to happen. Ironically, the
public awareness of high technology aroused by Jones
to challenge the existing infrastructure has helped
generate a profusion of high technology programmes in
the States so superficial that they pose no threat at
all to that infrastructure. Similarly, his declared
belief that 'in the area of high technology industry,
comparative advantage is not bestowed, rather
created' (Jones, 1983, page 7) seems to have fuelled
the assumption that comparative advantage can be
created instantly through high technology policy.

Venture Capital
The most significant high technology programme to be
implemented during Jones' brief responsibility for
the area was actually initiated - though only in the
most informal way - by the previous government. The
Australian Academy of Technological Sciences had
been approached in 1981 by the then Minister for
Science and Technology and asked for its advice on
the matter of venture capital. The Academy's
Committee, chaired by Frank Espie, was still
considering the matter when the Federal election was
called, precipitating the government into an early
release of a draft Espie Report and into
misunderstanding its recommendations (Ford, 1983).
The final Espie Report (Australian Academy of
Technological Sciences, 1983) was delivered to
Jones, who, with a history of success in using tax

concessions to stimulate the Australian film industry, agreed that similar treatment would work wonders for Australian high technology. The government rushed to select and license 10 management investment companies which would supply venture capital for high technology industry, encouraged by 100 per cent tax deductability for their own investors.

The Espie Report is a document of little rigour or intellectual merit (see Macdonald, 1983d). Its recommendations are largely dependent on its premises. It assumed - as was general before Jones - that the only ingredient missing from an otherwise perfect Australian high technology recipe was venture capital, the yeast to make technology rise. Certainly little venture capital market had developed in Australia, but almost no attention had been paid to the most obvious explanation - that there was little high technology in Australia worth financing. The explanation for that situation lies in an Australian infrastructure which is quite inadequate for high technology. Venture capital alone cannot compensate for such comprehensive inadequacy. Neither, of course, can government high technology policy, especially in the hands of those who have little appreciation of the requirements of high technology industry - particularly its information requirements - and who show scant regard for the importance of comparative advantage in high technology. The Australian States have little such appreciation and this is evident in much of their high technology policy.

High Technology Policies in the Australian States

Australia is a federation of six States and two Territories; collectively, with the Federal government, there are 14 parliaments - just about one for each million of population. Consequently, there is technology policy galore, and much of it unco-ordinated and conflicting (see Le Blanc, 1984; Australian Delegation to OECD, 1983). Table 8.1 gives some sort of appraisal of the programmes that exist, and shows where overlap and need for co-operation are greatest. Few of these programmes are specific to high technology, but many are new initiatives, at least at State level, and these tend to have been introduced with high technology as a main consideration. There is no point attempting to explore the maze of inspired high technology incentives here. Instead, it will be more

Table 8.1: Australian Federal/State Technology Related Assistance Measures

N/I = new initiative areas; X = programme exists

| | Federal | State | | | | | | | | Overlap | Need for co-operation (high, medium, or low priority) |
		NSW	Vic	Qld	WA	SA	Tas	ACT	NT		
Technology Development											
Basic research support	X						X		X	X	L
Industrial R&D grants (N/I)	X						X		X	X	H
Technology parks (N/I)		X		X	X	X		X		X	H
Government research organisations	X	X	X			X	X		X	X	L-M
Innovation centres (N/I)	X	X	X		X	X			X	X	H
Assistance to inventors (N/I)	X				X				X	X	M
Technology advisory centres (N/I)	X	X	X			X	X			X	H
Research associations	X										M
Patents	X										M
Technology Transfer											
Technology transfer organisations (N/I)	X	X	X	X	X	X		X		X	H
Technical information services (N/I)	X	X	X	X						X	H
Subsidies for technology consultants		X	X	X	X	X	X				M

242

Table 8.1: continued

| | Federal | State | | | | | | | | Overlap | Need for co-operation (high, medium, or low priority) |
		NSW	Vic	Qld	WA	SA	Tas	ACT	NT		
Industrial Development											
Expansion of establishment grants/subsidies		X	X	X	X	X	X				H
Relocation/removal assistance		X	X	X		X	X	X			H
Feasibility studies		X	X	X	X		X		X		H
Market research reports (N/I)		X	X	X	X		X		X		H
Purchasing preference	X	X	X	X	X	X	X			X	H
Offsets (N/I)	X	X	X							X	H
Advisory bodies to government (N/I)	X	X	X	X	X	X				X	M
Industrial Design Council	X	X	X	X	X					X	L
Productivity Promotion Council	X	X	X	X	X	X	X			X	L
International Issues											
Foreign investment	X										H
Technology transfer through joint ventures											
Export promotion (N/I)	X		X	X	X	X	X		X	X	H
Science and technology agreements	X		X		X					X	M
OECD	X									X	M

243

Table 8.1: continued

	Federal	State								Overlap	Need for co-operation (high, medium, or low priority)
		NSW	Vic	Qld	WA	SA	Tas	ACT	NT		
Social Aspects											
Studies of particular technologies	X	X	X		X	X				X	M
Consultative mechanisms (N/I)	X	X	X	X					X	X	H
Education/awareness raising (N/I)	X	X	X	X	X	X				X	H
Monitoring effects of technological change overseas	X	X	X	X	X					X	M
Government advisory bodies on technological change	X	X	X	X	X	X	X			X	H
Training grants etc.	X	X	X	X	X		X	X		X	H
Taxation Policy											
Income tax deductions for Industrial R&D	X										M
Payroll tax rebates		X	X								L–M
Investment allowance	X					X		X	X		L
Accelerated depreciation	X										L
Income tax deductions for equity investments in venture capital companies (N/I)	X										–

Table 8.1: continued

	State										Need for co-operation (high, medium, or low priority)
	Federal	NSW	Vic	Qld	WA	SA	Tas	ACT	NT	Overlap	
Finance											
Government development banks/corporations (N/I)	X	X	X		X	X	X		X	X	H
Private venture capital incentives (N/I)	X										H
Loan guarantees		X	X	X	X	X	X	X	X		H

Note: Table 8.1 is the list as supplied to Ministers. It contains omissions and inaccuracies which themselves suggest need for greater co-operation.

Source: Australian Department of Science and Technology Draft Discussion paper on Commonwealth/State Responsibilities for Technology, provided to Industry and Technology Ministers' Meeting, Hobart, June 1984, Attachment I.

constructive to consider those areas in which State programmes specifically tailored to high technology are concentrated, and then to attempt judgement of the overall impact of these policies on the development of high technology in Australia.

Financial incentives
Most States are anxious to offer companies a variety of financial incentives to locate within their boundaries. Typically these include investment allowances, loan guarantees, tax rebates and specific purpose grants. Such incentives are often still geared to the support of traditional manufacturing industry, and high technology firms which do not make a tangible product may not be eligible for support. Some schemes were originally designed to encourage the decentralisation of industry in what is actually a highly-urbanised country, but there is almost no interest now in using them for this purpose. Industry is very welcome wherever it chooses to locate within a State.

High technology industry is assumed to have difficulty raising funds from traditional financial sources, and various State agencies exist to fill this perceived gap in the market. For example, the Victorian Economic Development Corporation will lend to small firms with insufficient security to borrow from banks. However, the Corporation lacks technical expertise, is anxious to fund only projects at the post-prototype stage, does not encourage co-operation with non-Victorian enterprise, and is not supposed to lend to firms in industries receiving more than 26 per cent effective tariff protection. The argument for the last is that such firms have already dipped into the public purse, though such a provision would exclude high technology firms in heavily protected industries.

New South Wales runs a merit competition for high technology firms with $5 million available in loans and grants and a further $5 million in loan guarantees (which allowed advertisements in 1983 to promote the scheme by proclaiming that $10 million was being given away to high technology). There is an exhaustive and expensive selection process involving consultants, committees, and an application form deliberately designed to deter the least serious applicants. Four months after the first grants were announced by the responsible Minister, the winning firms had still received no cash. As cash flow problems prompted many firms to apply for grants, the scheme may actually have added to their burdens.

The Australian Capital Territory also has a scheme to give grants to firms and 80 per cent of its $750,000 budget for 1983 went to 'technology-based' companies. The unit which administers the grants has no technically-competent staff. The same unit also offers as an incentive to high technology firms its services in coping with the paperwork demanded by another unit responsible for planning high technology development. Both units are in the same government department. That department was responsible for bargaining with the American semiconductor firm that might have located in Canberra; it did not realise until years later that the planned fabrication plant would not produce chips ready for immediate Australian use and that assembly in Singapore would be necessary.

Western Australia, the first State to establish a second board appropriate for the listing of high technology shares, has a rather more entrepreneurial system. Anxious to receive a Federally-licensed MIC, it organised Westintech by gathering $10,000 from each of 70 founders, who also gave $500 to the campaign fund to lobby (successfully) for a licence. The State government contributed 10 per cent; the rest came from companies, educational institutions and even heads of government departments in their personal capacity. Westintech simply invests in likely high technology companies, the relevant management expertise coming from among its founders, who seem to hold poll positions in the Western Australian industrial network. South Australia failed to secure a Federal MIC in the first allocation and also instituted an alternative, though a more bureaucratic one. Its Enterprise Fund is a publicly-listed company with convertible notes guaranteed by the State government. The Fund is small, with no technical staff, and is unusual in that its responsibilities to its shareholders mean that it intends seeking the best return on its investments. Consequently, those investments will not necessarily be in South Australia.

Most States are sufficiently anxious to attract high technology and are sufficiently pragmatic to offer individual terms to likely firms. Queensland (where the Premier has publicly supported projects to power cars by hydrogen and to cure cancer) would seem to rely heavily on private negotiations and government-guaranteed loans to attract high technology firms. The Australian Capital Territory (ACT) claims that Federal supervision is too strict

to permit such practices and that it is, therefore,
at a disadvantage compared with the other States. Yet
it was the ACT which started the search for high
technology firms in 1980 by persuading the Federal
government to provide incentives worth nearly $20
million for its American semiconductor firm
(Macdonald, 1982). That extraordinary arrangement
came to nothing and Australia was spared a totally
inappropriate fabrication facility, but it set an
unfortunate precedent for States in their use of
financial incentives to attract high technology.

Innovation centres

There are several innovation centres in Australia and
more are planned. They provide assistance with the
commercial aspects of introducing new products and
processes. In South Australia, the centre has a small
budget with which to give even smaller grants, but
the main purpose of innovation centres is to give
advice to small businessmen on patenting, technical
matters, preparing business plans, and on market
prospects. With a tiny staff, the centre relies
heavily on outside consultants, some of whom are
sufficiently enthusiastic to contribute their
services free. Philosophies differ about whether
customers should be charged; in New South Wales a
flat fee is exacted, in South Australia the service
is free to clients, in Victoria, the Centre for
Innovation Development (one of two innovation
centres in the State) takes what it can get. Most
clients of the centres in New South Wales and South
Australia are individual inventors, a group badly in
need of advice in Australia, but not one with
momentous commercial prospects, or with much in
common with high technology (Macdonald, 1983e).
Nevertheless, both the South Australian centre and
the Product Innovation Centre in Western Australia
are located in technology parks. The Victorian centre
is highly selective and is determined to make money
from the ventures of its chosen clients. It has
little alternative for, unlike the other centres,
which are government funded and sometimes uncertain
whether they are intended even to cover their own
costs, it must support itself. Consequently, there is
little time to spare for individual inventors.

A largely separate function of innovation
centres is the provision of compact premises, often
with central services, appropriate for small high
technology firms. At a cost of $3½ million, South
Australia has provided such a building on its

technology park. The building is now fully occupied by tenants and construction of another is planned: the technology park is still almost empty. Such structures are planned in both Western Australia and Queensland, though Queensland has announced plans for bigger and better versions of nearly every high technology incentive introduced by the other States and little has yet been achieved.

High technology advisory units

All States have a Minister and department specifically responsible for technology, though technology often has some odd departmental bedfellows - small business in Queensland, decentralisation in New South Wales, and education in South Australia. Most States also possess special advisory units to provide information on high technology to government itself, to industry, or to the public at large. In serving industry, such units are presented with a reasonably straight-forward task: they explain incentives, they exhort and they cajole. In New South Wales, a major role of the Advanced Technology Unit is the administration of applications for the State's $10 million high technology incentive scheme. In serving government, the task is also fairly clear, given the unbridled enthusiasm of State governments for high technology: they provide evidence of how well the State is doing and of how well it can do. Where units have a responsibility to inform the public, though - or another ministry altogether - difficulties can arise. The Technology Research Unit in New South Wales and the Technology Advisory Unit in South Australia were both established in the days when new technology was regarded with political and popular suspicion, when it was seen as a threat to employment levels and working conditions. Such units were to be watchdogs, guardians of the public welfare. This role they continue, but now with massive difficulty in a climate that will brook no criticism of high technology, and certainly not of high technology policy. These units labour under considerable difficulties and their survival in such a hostile environment would seem unlikely.

Sometimes technology advisory units have had a part to play in formulating technology strategies for the States. In most cases there is actually a printed document announcing itself as just such a strategy, though high technology may be subsumed within greater planning for a greater economic development (e.g.,

Government of Victoria, 1984). State technology
strategy, however, is not an infallible guide to
State high technology policy. Policy in Queensland
bears almost no relationship to the State's only
strategy, but then the strategy was devised by a
department which has no direct responsibility for its
implementation (Premier's Department, 1983). In
Western Australia, the strategy was developed by
external consultants, but only after the major
programmes of policy had long been decided and
initiated (Falk, Johnston and Capp, 1984).

State governments have generally been anxious
to fill at least some of their new high technology
posts with recruits from the private sector. They
have argued, with some justification, that high
technology industry can be best advised and
administered by those who have some experience of the
activity. It is, however, not easy to lure successful
high technology entrepreneurs from private
enterprise, and those who succumb seem unable to
endure for long the distinctive style and pace of
work in the public service. Those who do are not
necessarily as adept at taking action in the public
interest as they were at taking action in their own
private interest.

Technology parks
The most prominent high technology initiative taken
by the States in Australia is the technology park.
All States except Tasmania and Victoria have at least
one, although the Victorian Department of Planning
seems to be unilaterally anxious to turn some spare
land above Flinders Street railway station into a
technology park. Precisely how many there are in the
whole country is uncertain (see Table 8.2); if the
definition of high technology is unclear, the
definition of a technology park (or science park or
research park) is even more so. Some private
developers have been quick to cash in on the cachet
and, without any government imprimatur, have re-
christened existing industrial estates. There have
even been advertisements offering 'high technology
warehouses'.

Technology parks are fascinating; they are high
technology as seen by planners and architects who
have transmogrified something strange and intangible
into something they can handle. They are high
technology as seen by politicians and bureaucrats who
wish to translate the nebulous into something
concrete. They generally offer a pretty landscape,

Table 8.2: Technology Parks in Australia

	Operating	Planned
New South Wales	–	2[a]
Queensland	2	2
Australian Capital Territory	1	–
Western Australia	1	–
South Australia	1	–
Victoria	–	–
Tasmania	–	–
Northern Territory	–	–

Note: a. plus 3 'technology centres' at Wollongong, Armidale and Newcastle

space-age buildings, potential agglomeration of firms, insulation from nasty industrial activity, a regulated environment, and - above all - the proximity of a university whence vital information is supposed to flow. Growing evidence from other parts of the globe that such factors are neither sufficient nor sometimes even necessary for the development of high technology industry, and that technology parks do not produce instant high technology industry (see Macdonald, 1985), have not been allowed to dampen enthusiasm for them in the Australian States. If there are obvious economic benefits to be reaped from high technology, there are equally obvious locational requirements for high technology and the technology park is seen as a neat embodiment of most of them.

The first technology park was established in Adelaide and some $3 million has been spent on land, site preparation and landscaping. In 1984 there was only one tenant on site - building his own factory, having been lured from Canberra by superior incentives. The technology park is alongside the South Australian Institute of Technology, some 13 kilometres from the relative civilisation of Adelaide. There is, though, the innovation centre and its computer links with the Institute of Technology do represent contact of a sort. Land adjoining the park is cheaper than park land and development there would avoid many of the regulations to be imposed on park residents. The Canberra technology park is managed by a private company rather than by public servants, and one which is refreshing in its realism towards high technology. While the local public servants claim that Canberra has absolutely everything a high technology enterprise could possibly desire, the property developers see the only real advantage to be proximity to the bureaucrats and order books of Federal government. Quite sensibly, the developers seek to entice only sizeable firms of some standing to the park, and wish to avoid wasting their resources on marginal firms likely to fail. The incentive scheme run by the Canberra Development Board provides the developers with a convenient screening mechanism - any high technology firm which needs a grant from the Canberra Development Board is likely to be a marginal firm and is unlikely to be taken seriously by the developers.

In Perth the technology park is alongside the Western Australian Institute of Technology. The original idea - as elsewhere - came from ministerial level and was immediately translated into a verdant map by a local landscape architect. Only after the decision was taken to proceed with the technology park was a consultancy study launched to assess its viability. Having received questionnaire responses from 12 firms, the study was completed in May 1982. The only other assessments of the technology park were those undertaken for an MBA thesis, and by some students at a local College of Advanced Education. In New South Wales, the government was committed to creating a technology park at Homebush Bay in Sydney by the Premier in 1984. What assessment there is has come not from the Premier's Department, but from the Department of Industrial Development and Decentralisation, which has declared that 'the description of the area as a potential "advanced technology park" goes against all conventional wisdom on technology parks' (James, 1985, page 7). There would appear to be some justification for that statement: Homebush Bay must be one of the least attractive places in Sydney; it is reclaimed land with poor communications and subject to noxious odours from the abbatoirs which also occupy the site. While pretty surroundings do not ensure success in high technology, positively ugly ones make no contribution whatsoever. As for the three 'technology centres' proposed for provincial New South Wales, it is hard to see what comparative advantage Armidale has as a biotechnology centre or Wollongong as a robotics centre. These places, and the third - Newcastle - have only a desperate need for any sort of economic activity.

Queensland is to have more technology parks than any other State, though there is some uncertainty about where they are or might be. In Queensland the distinction between a technology or science or research park and any other industrial development

seems to be especially blurred. As the responsible
minister said recently in announcing plans for one
science park and the existence of two technology
parks:

> We are concentrating on technology at this time
> because we have the tools and we have the need
> to apply them but we don't necessarily know
> where or how to apply them. (Ahern, 1985, page
> 1)

Overall impact

High technology policy is still new to Australia and
it is really far too soon to judge the success of the
many programmes intended to stimulate the activity.
However, certain characteristics of policies and
programmes are already evident and they do not
generally bode well for the development of high
technology industry. Despite the existence of an
Industry and Technology Ministers' Council to coordinate the various State and Federal initiatives in
this area, it is evident that there is almost no co-
ordination, that States introduce high technology
measures not simply in ignorance of measures taken by
other States or by the Federal government, but in
defiance of them. The Australian States have a long
and unhealthy tradition of rivalry with the Federal
government and of competition with each other, a
tradition which sits ill with the heavy involvement
of government in the regulation of the economy. The
result - ironically - is intense competition in
methods of industry protection, and in that sort of
competition there are likely to be no winners.

It is common in Australia for high technology
firms to play one State against another, or against
the Federal government; to use incentives offered in
one place in order to bargain for better incentives
in another. Some programmes almost encourage such
behaviour. For instance, the Federal government
operates an offsets scheme largely to improve the
technological capacity of industry in Australia.
Foreign companies selling to the Federal government
are expected to offset 30 per cent of the value of
the contract, typically by producing part of the
order in Australia. That activity must take place
somewhere and States compete vigorously to attract
it; so vigorously, in fact, that what was a penalty
imposed on the overseas firm at Federal level can be
turned into a reward for the firm at State level. The
Federal offsets scheme is not well run and it is

difficult to determine just what technological improvement has resulted from its attempts to allocate its vast resources (Committee of Review on Offsets, 1985). This has not deterred some States from running their own offsets schemes in complete disregard of comparative disadvantage. Unsuccessful attempts have recently been made to abandon State industry preference in State government procurement because of the fragmentation of industrial activity it encourages. There seems to be little awareness that the same fragmentation is likely to arise in high technology industry, brought about by State incentives in general, and by State offsets schemes in particular.

During the tenure of Barry Jones as Minister for Science and Technology, a Federal Technology Strategy was produced in draft (Department of Science and Technology, 1984b); that has been slow to find final form. Meanwhile, most of the States have their declared technology strategies, each paying scant regard to the others. At best, effort will be duplicated and resources wasted: at worst, plans will be confounded and hopes frustrated. Take just a single example of the typical conflict: the Federal government insists that applicants for its Assistance to Inventors Scheme must first apply for a patent; in South Australia, the innovation centre (receiving Federal support) advises inventors (who could also be receiving Federal support) that they should not be rushed into applying for a patent. The example is trivial, but it exemplifies the confusion that now emanates from government high technology policy in Australia.

This unhappy situation has arisen largely because inadequate attention has been paid to the matter of comparative advantage in high technology. It has simply been assumed by the Federal government and by the individual States that whatever they have to offer is conducive to the prosperity of any high technology activity, and that any high technology firm will thrive in such conditions no matter what it does. Such assumptions for firms in established industries would be clearly untenable, but high technology is new, it has a mystique that does not encourage mundane calculations of comparative advantage. It promises inestimable benefits in return for little more than faith and political will. Of course, it is argued that market forces will also influence high technology firms and will rapidly deal with those firms that are not viable. However, government policies are likely to have added so many

distortions to an already imperfect market that its
capacity to operate may now be severely impaired.
Anyway, market forces have long been powerless
against government support in dealing with the many
unviable firms in other Australian industries. In
fact, much of the aura of infallibility which
surrounds high technology in Australia has actually
been generated by government policies, which are then
encouraged by the very aura they have created. The
Federal government, and many of the States, now issue
glossy periodicals praising the virtues of
technology in general and high technology in
particular. Even independent science and technology
advisory councils - of which there are several -
bring forth advice on such matters that is more
instinctive than reasoned in its optimism (e.g.,
N.S.W. Science and Technology Council, 1983. See
Lamberton, 1985). Assumption of comparative
advantage in high technology stems from a fundamental
lack of understanding of high technology industry;
the readiness of this assumption springs from the
profound belief in Australia that the costs of high
technology are negligible and the benefits immense.

Policy Achievements
Government high technology policy in Australia is
still so very recent that it is really far too soon
to assess its achievements. Policy may eventually
produce much that has not yet happened. There have
already been some developments, such as the growth in
the supply of venture capital, but it is impossible
to be sure what would have occurred in the absence ot
policy. Other developments, such as agglomeration of
high technology firms, are noticeably lacking, and
these might well have ocurred more readily in the
absence of policy. There are high technology firms in
Australia, whatever definition of the term is
employed, but whether they exist because of, or in
spite of, policy is uncertain.
    What is certain is that high technology firms
suffer the same difficulties that beset other firms
in Australia - the cumbrous tax system, the tiny
domestic market, the distances between urban areas,
for example - without being part of the established
industrial structure. High technology firms are
involved in new activities which do not fit neatly
into the present pattern of interests, assistance and
support. This, of course, is one justification of
incentives for high technology industry, but such
policy is clearly a second best solution. Ideally,

the assistance afforded established industrial activity should be reduced so that high technology industry is not disadvantaged, but that is not the Australian way. High technology industry must receive its entitlement of support - and that requires specific policy - if it is to compete for resources against established industries. But if the most important benefit from high technology activities is the impetus and example they can give for the restructuring of the economy, much of this is likely to be negated by high technology policy designed to leave the existing structure untouched.

It is not the level of support for high technology that is exceptional in Australia, but the fervour of that support. Enthusiasm has been necessary to achieve high technology policy, but that same enthusiasm has discouraged both rigorous assessment of comparative advantage and appreciation of the fundamental information characteristics of high technology industry. Australian high technology policy is most notable for its neglect of both these matters. More effective policy demands, of course, co-ordination by the several Australian governments, but also a consensus on what is possible and why it should be desirable. There is none at the moment. Effective policy demands targeting and probably a concentration on niche markets overseas. It demands continual assessment of Australia's competitive position in high technology and acknowledgement that access to, and skill in handling, information are likely to be critical in determining this. There must also be recognition that little is likely to be achieved in the short term and that benefits in terms of a restructured economy will not be evident in high technology industry itself. But even wise and considered high technology policy runs the risk of distorting market forces so much that they become unrecognisable, and of stifling entrepreneurial vigour. Even if Australia eventually achieves wise and considered policy for the development of high technology industry, that problem will remain to be faced.

References
Ahern, M. (1985) 'Technology and politics - a necessary partnership for future economic growth in Queensland', paper presented to National Party symposium (Brisbane) April
Armington, C., Harris, C. and Odle, M. (1983) Formation and Growth in High Technology Businessess:

A Regional Assessment, Brookings Institution, Washington D.C., September, Appendix A

Australian Academy of Technological Sciences (1983) Developing High Technology Enterprises for Australia, Parkville

Australian Computer Equipment Manufacturers' Association (1983) Submission to IAC Inquiry into Computer Hardware and Software, Melbourne, April

Australian Delegation to OECD (1983) 'Technology policies and programs in Australia', paper presented to OECD workshop 'Research, Technology and Regional Policy', Paris, October

Australian Industries Development Association (1974) Industrial Research and Development. The Case for Government Assistance, Canberra

Australian Science and Technology Council (1978) Science and Technology in Australia 1977-78, AGPS, Canberra, Vol. 1A, 44

Ballinger L., Hope K. and Utterback J. (1983) 'A review of literature and hypotheses on new technology-based firms', Research Policy, 12, 1-14

Baram M. (1968) 'Trade secrets: what price loyalty?' Harvard Business Review, 46, (6), 66-74

Begley S. and Carey J. (1984) 'Toxic trouble in Silicon Valley', Newsweek, 7 May 85

Beyers B. (1983) 'The groves of silicon', Enterprise, August, 11-14

Board of Trustees, Stanford University (1974) Stanford University Land Use Policies, Stanford University, 12 March

Braun E. and Macdonald S. (1982) Revolution in Miniature. The History and Impact of Semiconductor Electronics, Cambridge University Press, Cambridge

Carlson E. (1984) 'Listing top high-tech States depends a lot on definition,' Wall Street Journal, 3 January

Cetron M. and Davison H. (1977) Industrial Technology Transfer, Noordhoff, Leyden, 257-74

Committee of Review on Offsets (1985) Report, AGPS, Canberra

Commonwealth Record (1982) 'Development of special purpose silicone chips', 23-29 August, 1173

Department of Science and Technology (1981) Creating High Technology Enterprises, AGPS, Canberra

Department of Science and Technology (1984a) National Technology Conference Proceedings, AGPS, Canberra

Department of Science and Technology (1984b) National Technology Strategy. Discussion Draft, Canberra, April

Dianda M. (1984) 'The Silicon Valley being reshaped

by Silicon sprawl' <u>Peninsula Times Tribune</u>, 22 April, Al-A8

Dorfman, N. (1983) 'Route 128: the development of a regional high technology economy', <u>Research Policy</u>, 12, 299-316

Durie J. (1981) 'High technology claims government has abandoned it', <u>Financial Review</u>, 12, August

<u>Economist</u> (1983) 'How to cut unemployment', 28 May, 15

<u>Economist</u> (1984) 'Europe's technology gap', 24 November, 99-110

Encel S. (1970) 'Science, discovery and innovation: an Australian case history', <u>International Sociology of Sciences Journal</u>, 22, (1), 42-53

Falk, J. Johnston R. and Capp D. (1984) <u>Framing a Technology Strategy for Western Australia</u>, Centre for Technology and Social Change, University of Wollongong, July

Feller I (1984) 'Political and administrative aspects of state high technology programs', <u>Policy Studies Review</u>, 3 (3-4), 460-6

Ford J. (1983) 'Anger over technology "muddle",' <u>Australian</u>, 25 February, 7

Fraser M. (1983) (Prime Minister) address to Young Liberals National Convention, 14 January 8-9

Gannicott K. (1980a) 'Research and development incentives', in Committee of Enquiry into Technological Change in Australia, <u>Technological Change in Australia</u>, AGPS, Canberra, 4, 287-314

Gannicott K. (1980b) 'Simple economics and difficult policies: the case of public money for research and development', paper presented to 50th ANZAAS Congress, Adelaide

Gannicott, K. (1980c) 'The balance of R&D in Australia: a contary view', <u>Search</u>, 11 (12), 406-9

Gibbons M. and Johnston R. (1974) 'The roles of science in technological innovation,' <u>Research Policy</u>, 3, 220-42

Green R. and Ronayne J. (1980) 'Innovation policy and the balance of R&D in Australia,' <u>Search</u>, 11 (12), 403-6

Green R. and Ronayne J. (1981) 'How much use is R&D', <u>Search</u> 12 (7), 203

Government of Victoria (1984) <u>The Next Step. Economic Initiatives and Opportunities for the 1980s</u>, Melbourne, April, 110-18

Gregerman A. (1984) 'Competitive advantage - framing a strategy to support high-growth firms,' in <u>Commentary of the National Council for Urban Economic Development</u>, Washington D.C., Summer, 18-23

Heywood G. (1983) 'Fraser's high-tech switch,'

Financial Review, 27, January 1
Hodgman M. (1981) Hansard, Representatives, 26 March
Huyer H. (1974) 'The future of the electronics
industry in Australia,' paper presented to the
Institution of Radio and Electronics Engineers
Australia, Sydney, 27 September
James R (1985) 'Homebush Bay Advanced Technology Park
- an assessment', working paper, Department of
Industrial Development and Decentralisation, Sydney,
January
Jones B. (1982) Sleepers Wake! Technology and the
Future of Work, Oxford University Press, Melbourne
Jones B. (1983) 'Keynote address', Management
Technology Education Conference on Sunrise
Industries, Sydney, 31 May, 7
Kutzmann D. (1984) 'Valley's congestion chases job-
seekers away', San Jose Mercury, 18 March
LaDou J. (1984) 'The not-so-clean business of making
chips', Technology Review, May-June, 23-8
Lamberton D. (1984) 'Australia as an information
society - who calls the shots?', in Department of
Science and Technology, National Technology
Conference, AGPS, Canberra, 125-31
Lamberton D. (1985) 'The ASTEC Technological Change
Committee and science and technology policy,'
Search, 16 (3-4), 59-60
Lamberton D., Macdonald S. and Mandeville T. (1982)
'Productivity and technological change: towards an
alternative to the Myer's hypothesis', Canberra
Bulletin of Public Administration, 9 (2), 23-30
Le Blanc M. (1984) A Complete Guide to Technology
Assistance in Australia and New Zealand, Rydge
Publications, Sydney
Lynch P. (1981) Hansard, Representatives, 31 March,
1981
Macdonald S. (1979) 'The need to succeed,' Journal of
General Management 4 (3), 74-83
Macdonald S. (1982) 'The relevance of Silicon Valley
to Australia', Australian Electronic Engineering, 15
(3) 38-46
Macdonald, S. (1982-3), 'Faith, hope and disparity.
An example of the public justification of public
research', Search, 13 (11-12), 290-9
Macdonald S. (1983a) 'Technology beyond machines,'
in S. Macdonald, D. Lamberton and T. Mandeville,
(eds) The Trouble with Technology, Frances Pinter,
London, 26-36
Macdonald S. (1983b) 'The lowdown on high technology
industry in Australia', in A. Birch (ed) Science
Research in Australia: Who Benefits?, Centre for
Continuing Education, Australian National

University, Canberra, 153-72

Macdonald S. (1983c) 'Agricultural improvement and the neglected labourer,' Agricultural History Review, 31 (2), 81-90

Macdonald S. (1983d) Book Review, Prometheus, 1(2), 389-91

Macdonald S. (1983e) 'Australia - the patent system and the individual inventor,' European Intellectual Property Review, 5 (6), 154-9

Macdonald S. (1985) 'Towards higher high technology policy,' paper presented to joint OECD/Italian seminar 'Opportunities for Urban Economic Development,' Venice, June

Macdonald S. and Mandeville T. (1982) 'Sources of technological information in Australia,' paper delivered to 52nd ANZAAS Congress, Sydney

Mandeville T. and Macdonald S. (1983) 'Information technology and employment levels,' in S. Macdonald, D. Lamberton and T. Mandeville (eds) The Trouble with Technology, Frances Pinter, London, 169-77

Mandeville T., Macdonald S., Thompson B. and Lamberton D. (1983) Technology, Employment and the Queensland Information Economy, report to the Queensland Department of Employment and Labour Relations, Department of Economics, University of Queensland, October

Markusen A. (1983) High Tech Jobs, Markets and Economic Development Prospects, Working Paper No. 403, Institute of Urban and Regional Development, University of California, Berkeley, April

NSW Science and Technology Council (1983) Development of High/New Technology Companies in New South Wales, Sydney, October

OECD (1981) Information Activities, Electronics and Telecommunications Technologies, Paris

Office of Technology Assessment (1984) Technology, Innovation and Regional Economic Development, US Congress, Washington DC, July

Ord Minnet Ltd (1985) Australian Technology and Related Venture Capital Company Review, Sydney, 15 March

Porat M. (1978) 'Communication policy in an information society' in G. Robinson (ed) Communications for Tomorrow - Policy Perspectives for the 1980s, Praeger, New York

Premier's Department, Queensland (1983) Towards a Strategy for Technological Development in Queensland, Brisbane, October

Premus R. (1983) Location of High Technology Firms and Regional Economic Development, Joint Economic Committee, US Congress, US Government Printing

Office, Washington DC

Rattigan G. (1977) 'Comparison of the IAC and Jackson Committee approaches to industrial development,' Australian Economic Papers, 16 (28), 26-43

Riche R., Hecker D. and Burgan J. (1983) 'High technology today and tomorrow: a small slice of the employment pie,' Monthly Labour Review, 106 (11), 50-8

Rogers E. and Larsen J. (1984) Silicon Valley Fever, Basic Books, New York

Rudy R. (1984) 'Hunger on Peninsula a concern, experts say', Peninsula Times Tribune, 28 January, A1-A8

Rumberger R. and Levin H. (1984) Forecasting the Impact of New Technologies on the Future Job Market, Project Report No. 84-A4, School of Education, Stanford University, February

Science and Government Report (1983) 'Commentary: the science education stampede ... A dissenting view on computer 'revolution', 1 October, 6-7

Senate Standing Committee on Science and the Environment (1979) Industrial Research and Development in Australia, AGPS, Canberra, 175-7

Soiffer, B. (1984) 'Toxic suits pile up in Silicon Valley,' San Francisco Chronicle, 18 June, 2

Teece D. (1976) The Multinational Corporation and the Resource Cost of International Technology Transfer, Ballinger, Cambridge Mass. 23-30

Tegart W. (1983) 'Technology policy: the case for a more selective approach,' Ascent, 1 (1), 13-16

Thomson D. (1983) 'Official opening' in Department of Science and Technology Finance for Technology Ventures, AGSP, Canberra, 4

Tomaskovic-Devey D. and Miller S. (1983) 'Can high-tech provide the jobs?,' Challenge, 26 (2), 57-62

Weber D. (1981) 'What's cooking at Stanford?,' California Business April, 45-7, 74

Webster D. (1980) 'How an Australian computer system manufacturer can survive in the face of international competition,' paper delivered to DEC Users Symposium, Sydney, July

Wild J. (1984) 'High technology - is it the answer?', Australian Director, June/July, 47-8, 51-2

Windschuttle K. (1984) 'High tech and jobs,' Australian Society, 3 (5), 11-13

Chapter Nine

THE DEVELOPMENT OF HIGH TECHNOLOGY INDUSTRY IN JAPAN

Hisao Nishioka and Atsuhiko Takeuchi

Introduction

It is widely accepted in Japan that the term 'high technology industry' implies microelectronics, mechatronics, new materials, new energies, optical fibre and communication apparatus, biotechnology, new industrial and/or community systems and so on. However, the machinery and equipment industry (hereafter, the 'machinery industry') which consists of industrial machinery, electrical machinery, transport and equipment and precision instruments, or a little more widely, the 'metal processing' (or fabricating and assembling) industry, which consists mostly of the machinery industry and partly the fabricated metals industry, are more frequently characterised as high technology sectors. This is for two reasons. First, the machine industry has typically and effectively embodied high technology in its products and/or processes. Second, most other industries, in which important innovations can be found, have so far performed less well than the machinery and metal processing industries. (1)

The machinery industry usually has complicated inter-, and particularly intra-industrial linkages, and in many cases develops through such linkages (e.g. Kawashima, 1964). This could be the reason why Weber (1909, pp. 195-6) indicated 'contact advantages,' especially in the machinery industry (Aoki, 1960, pp. 227-20). The traditional heavy chemical industry, using basic resources, generally undertakes large scale investments and can employ many workers. Compared to its own size of plant, however, the related and successive effects on regional development are often meagre. Even if the effects are significant, they might leak out from the region, and when newer competitors appear anywhere plants in this sector can rapidly lose their market.

The reverse situations can be the case in the machinery industry.

In Japan, the machine industry has developed quite rapidly, in particular since the second oil crisis. As is seen in Table 9.1, it occupied 36.2 per cent of Japan's total shipment values of all manufactures in 1983, and if fabricated metals are included, 41 per cent. Figure 9.1 shows that the value of shipments (adjusted by price indices) of the metal processing industry accounted for more than 45 per cent of the national total in 1981, and is expected to reach over half of the national total in the future. It should be noted that the electrical machinery sector alone produces twice the shipments-value of the iron and steel industry which led to the previous rapid economic development of Japan (Table 9.1).

Figure 9.1: Industrial Structure of Japan by Four Types of Manufacturing Based on Adjusted Shipments-Value (1976-82)

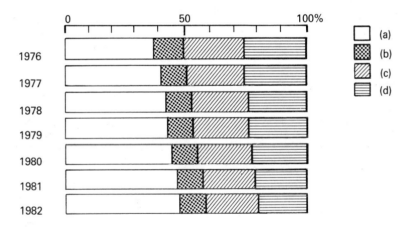

Note: Each industry type consists of several industry groups as indicated by numbers in the following. (The numbers are the same as those in the left hand of Table 9.1).
(a) Metal Processing Type: 15-19.
(b) General Merchandise Type: 3, 5, 7, 10, 11 and 20.
(c) Local Resources Type: 1, 2, 4 and 12.
(d) Basic Resources Type: 6, 8, 9, 13 and 14.

Source: JILC (Japan Industrial Location Centre), 1984, Unpublished Data of Shipments-Value.

Table 9.1: Shipments Value in Manufacturing by Industry Group (1983)

Industry Group	Value of Shipments (12 Yen)	Relative value in 1982 (1980 Values 100)	Shares by Industry (%)	
1. Food and Kindred Products	26.3	104.3	11.2	
2. Textiles	8.1	98.8	3.4	
3. Apparel and Related Products	3.3	102.1	1.4	
4. Lumber and Wood Products	4.2	96.3	1.8	
5. Furniture and Fixtures	2.9	98.8	1.2	
6. Pulp, Paper and Allied Products	7.1	102.3	3.0	
7. Publishing, Printing and Allied Prodts.	8.2	104.9	3.5	
8. Chemical and Allied Products	19.2	104.0	8.2	
9. Petroleum and Coal Products	14.2	90.1	6.1	
10. Rubber Products	2.7	107.1	1.2	
11. Tanned Leather Products	1.0	99.6	0.4	
12. Ceramic, Stone and Clay Products	8.7	100.7	3.7	
13. Iron and Steel	16.1	92.1	6.8	
14. Nonferrous Metals and Products	6.9	102.6	3.9	
15. Fabricated Metal Products	11.2	98.9	4.8	
16. Machinery, excl. Electric incl. Electric incl.	20.3	101.3	8.6 ⎫	⎫
17. Ordnance Accessories			⎬ 36.2	⎬ 41.0
18. Electric Machinery, Equip and Supplies	31.6	114.2	13.4	
19. Transport Equipment	29.6	103.2	12.6 ⎭	
20. Precision Equipment	8.3	104.7	1.6	⎭
Other	10.3	108.8	4.4	
Total	235.6	102.2	100.0	

Note: The parenthesised numbers at the left edge are used for convenience and different from those of two digits for intermediate industry group in the standard classification.

Source: MITI, 1984, 'Preliminary Report on the Census of Manufacture, 1983,' pp. 6–7.

Of course, the machinery industry incorporates many non-high technology products/processes, while other industries often possess or are developing high technology products/processes. Therefore, to analyse exactly the distribution or trends in high technology industry, the lowest levels of industrial classification should be considered, and accordingly, medicines and drugs for instance, which belong to the chemical industry, should be selected.

Whether software can be included with the manufacturing industry is another problem, but in section three of this paper both software and system houses are discussed since they are playing vital roles in high technology industry.

The next (second) section describes the distribution of metal processing types of industry and assesses new possible locations for selected high technology products. In the third section, the actual circumstances of high technology industry in the Tokyo metropolitan area and also three local areas of Hamamatsu, Suwa and Nagaoka, are considered.

The spatial units adopted in the next section are the 14 regions used by the MITI (Ministry of International Trade and Industry - see Figure 9.2 (2)). Different divisions are used in the third section for a different purpose, though many regions are identical such as Coastal Kanto in the former and southern Kanto (or Tokyo metropolitan area) in the latter (see Figure 9.6).

The third section is written by Takeuchi, and the remainder by Nishioka.

Distribution and Future Locational Trends of High Technology Industry and All Manufacture

1.   Changes of Industrial Areas
(a)   **Changes between Pre- and Postwar Periods**. Prewar Japanese industry developed heavily in the four largest industrial areas: (1) Osaka-Kobe, (2) Tokyo-Yokohama, (3) northern Kyushu and (4) Chukyo (i.e. Nagoya and its surroundings) in that order. During the war, Tokyo-Yokohama and Chukyo began to compete respectively with Osaka-Kobe and north Kyushu. Since the war their status have been reversed, and this, together with other factors, has caused the 'East-West reversal' (Nishioka, 1962, 1963a Chs. 6-7, 1963b, 1966, 1981; Nishioka and Oshiro 1979).

Comparing the regional shares of national output in the first half of the 1950s to those in 1935, northern Kyushu, which was primarily coal

mining iron and steel producing and an area depending
upon Chinese resources, declined. Osaka-Kobe, which
depended heavily on textile production and trade,
also followed the same pattern. Chukyo steadily
advanced with metal products, petrochemicals, and
later automobiles. Tokyo-Yokohama surpassed Osaka-
Kobe and assumed the prime position among the major
industrial areas, with metals, machinery and
chemicals. Even in agriculture, the growth rate in
eastern Japan (especially in Tohoku, the prewar
typical backward region) surpassed western Japan.
The interprefectural per capita income differentials
largely lessened in the postwar period, but eastern
Japan gained dominance in this period. Northern
Kyushu eventually dropped from the fourth position in
industrial production.

After World War II, Japan intended to develop
the heavy and chemical industries which were believed
to be necessary to restructure her economy, as well
as to avoid the possible future frictions with newly
industrialising countries. In the middle 1960s the
proportion of heavy and chemical industry relative to
total industry in Japan reached the level of western
advanced countries, but since then Japan has become
aware of the importance of the machinery or metal
processing industry. By the time Osaka reinforced its
steel production, the significance of basic material
industry for regional development (particularly in
such an advanced area as Osaka) already had become
doubtful.

Industrial activities have gradually extended
from established centres to, and become more vigorous
in, outer zones. The names of the three largest
industrial (or rather metropolitan) areas of Tokyo,
Osaka and Nagoya have become familiar and the
industrialised areas have been enlarged. For
example, Tokyo-Yokohama area was previously
considered as both prefectures of Tokyo and Kanagawa,
and later Chiba Prefecture was added, sometimes under
the name of Tokyo-Yokohama-Chiba. Today the Tokyo
metropolitan area is usually taken to contain four
prefectures, including Saitama.

Specialists have divided the heavy and chemical
industry into 'metal processing' and 'basic
materials or resources'. A four-way classification
of industry including further 'local resources' and
'general merchandise' is frequently used (e.g.
Kawashima, 1980). (See note below Figure 9.1).

(b) **Structural Changes Since the Oil Crises.** Postwar
Japanese manufacturing industry has developed mainly
in the so-called Pacific Coast Belt Region, extending
from South Kanto to North Kyushu (Figure 9.2),
particularly around the three largest metropolitan
areas of Tokyo, Osaka and Nagoya. This concentration
is due to historical inertia, climate and topography,
existing   infrastructure,   domestic   and   overseas
markets and intra- and interindustrial linkages
including co-operating and subcontracting systems.
These already densely populated areas attracted more
people of working age from rural regions such as
Hokkaido,   Tohoku,   Shikoku   and   Kyushu.   The
concentration of industry and population has caused
numerous problems, such as traffic congestion and
industrial   water   shortage,   and   aroused   public
attention to the need for an industrial location
policy (Nishioka and Oshiro, 1979).

Figure 9.2: Areal Divisons of Japan (employed
frequently by the MITI for locational statistics) and
the Pacific Coast Belt Region (roughly defined).

The concentration trend peaked in the late 1960s and serious pollution-related habitational problems were recognised as an important social problem. The Japanese government has restricted factory location in the major metropolises and improved the transportation network in the country since the late 1950s. The Industrial Relocation Promotion Act was enacted in 1972 for promoting industrial relocation from the major metropolises ('Industry Relocation Promotion Areas') to less-developed and/or underpopulated areas ('Induction Areas').

The oil crisis of 1973 has brought about a number of structural changes to Japanese industry and its location; (1) The annual growth rate of GNP, which was on average 8.9 per cent in 1969-1973, was -1.2 per cent in 1974 and 2.4 per cent in 1975, though it averaged 4.7 per cent over 1976-1984. (2) The indices of industrial production and shipments showed a sudden drop, though these recovered in 1977 and 1978 respectively. (3) The basic material industries such as iron and steel, chemicals, coal and petroleum products, paper and pulp and textile have declined or stagnated, while the machinery industry has steadily developed as mentioned earlier. Japan has so high a dependency of oil supply on overseas sources (99.8 per cent even in 1983) that she has made a great effort to save oil and reduce production costs. Thus, combined with other reasons, high technology machinery industry has developed. (4) New possible locations measured by the number and area of industrial sites acquired by firms (hereafter, 'new locations') have reduced to less than half of those before the first oil crisis (5,088 cases in 1973 to 1,273 in 1977 and 2,357 in 1984). However, the machinery firms have shown a sharp increase in new locations after the second oil crisis occurred in 1978-79. (5) MITI, which so far had made much of the traditional heavy and chemical industry, turned in 1980 to advocate the idea of the development of Japanese industry on the basis of high technology, and suggested a 'Technopolis Concept' (For Technopolis and related arguments, see Appendix A and Nishioka, 1984, 1985).

We shall now discuss the tendencies of the existing and possible future distribution of high technology industry, comparing with those of all manufactures.

2. __Existing Distribution of All Manufacturers and Metal Processing Industry by Region__

Table 9.2 shows regional shares of Japan's industrial shipments-value in metal processing as well as in all manufactures in 1976 (a middle year between the first and second oil crises) and in 1982. It tell us that: (1) Coastal Kanto accounted for more than 26 per cent of all manufactures and about 30 per cent of metal processing, and if we can figure for both Coastal and Inland Kanto, all the Kanto region accounted for more than 35 per cent and 42 per cent of these sectors; (2) Coastal Kinki accounted for about 15 per cent of all industries and about 13 per cent in metal processing and fell behind Tokai; (3) both central areas (both Coastal Kanto and Kinki) reduced their shares in both industrial cases, but their neighbours (both Inland Kanto and Kinki, especially Inland Kanto), increased in both cases; (4) Tokai steadily

Figure 9.3: Industrial Structure of Shipments-Value by Four Types of Manufacturing in Each Region (1976 and 1982)

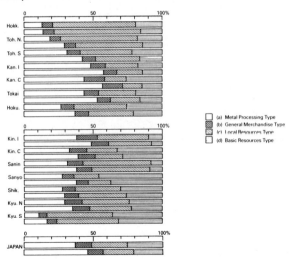

(a) Metal Processing Type (b) General Merchandise Type (c) Local Resources Type (d) Basic Resources Type

Note: In each region, the above graph is for 1976 and the below, for 1982.

Source: JILC, 1984, Unpublished Data.

Table 9.2: Regional Shares of Adjusted Shipments Value of All Manufacturers and Metal Processing Industry (1976 and 1982)

Region		All Manufactures			Metal Processing		
		1976 (%)	1982 (%)	Average Annual Increase (1982/76, %)	1976	1982	Average Annual Increase (1982/76, %)
Hokkaido		2.47	2.05	1.72	0.88	0.60	2.23
Tohoku	N.	1.33	1.29	4.41	0.63	0.82	12.24
	S.	4.28	4.37	5.33	3.60	4.00	10.79
Kanto	I.	8.30	9.87	8.01	10.76	12.38	11.40
	C.	26.76	26.48	4.77	31.49	29.93	7.91
Tokai		17.09	18.61	6.46	20.01	21.56	10.19
Hokuriku		2.38	2.37	4.87	1.75	1.91	10.40
Kinki	I.	3.71	4.17	7.05	3.83	4.4	11.43
	C.	16.29	14.81	3.30	14.56	12.87	6.61
San'in		0.56	0.55	4.81	0.47	0.47	8.26
San'yo		7.70	6.84	2.91	5.84	5.60	8.08
Shikoku		3.02	2.66	2.72	2.26	1.69	3.65
Kyushu	N.	4.19	4.06	4.39	3.34	3.10	7.50
	S.	1.90	1.86	4.53	0.53	0.67	13.02
Japan Total	(12 Yen)	100.00	100.00	4.95 (Av.)	100.00	100.00	8.83 (Av.)
Value		137.8	184.2		50.9	84.6	

Note: Underlined ratios are larger than the national average.

Source: JILC, (1984) Unpublished Data.

grew in both cases and in the field of metal processing exceeded Coastal Kinki; (5) among others, South Tohoku alone improved the shares in both cases, and North Tohoku, Hokuriku and South Kyushu improved in metal processing; all of these regions are characterised by relatively easier access to the Tokyo area (in the case of South Kyushu, through air transport) and cheaper labour and land; and (6) within western Japan, only Inland Kinki (in both) and South Kyushu (in metal processing) expanded their shares.

Figure 9.3 indicates the industrial structure of shipments-value by the above mentioned four industrial categories in each region in 1976 and 1982. Clearly, the figure shows that the share of metal processing increased in every region, but only four contiguous regions (both Inland and Coastal Kanto, Tokai and Inland Kinki) exceeded the national average in 1982 of 45.91 per cent, and Inland Kinki alone was within western Japan.

In conclusion, Coastal Kanto (Tokyo metropolitan area) is the largest centre of industry (especially metal processing) from which metal processing is diffusing, in particular to neighbouring and contiguous regions.

3. **Locational trend in the near future: Shipments-Value vs. New Locations and All Manufacturers vs. High Technology Industry**

(a) **Shipments-Value and New Locations of All Manufacturers.** Now we shall compare the changes of regional shares in shipments-value of all manufactures, shown already on Table 9.2, with those in new industrial sites. On the left half of Table 9.3 are given the regional shares in total numbers of new sites over six years, including the second oil crisis (1976-1981) and also during three recent years (1982-84). On the right half, a plus sign signifies an increase in regional share and a minus sign a decrease, either of the shipment-value (in the first column on the right half) or of the new sites (in the second). (1) Both centres (Coastal Kanto and Kinki) decrease between the two periods in both shipments and new sites. This reflects the effects of restrictive laws, higher wages, and especially expensive land. (2) Their neighbours (Inland Kanto, Tokai, Inland Kinki and San'yo) show different situations. The closer is a region to Tokyo, the

Table 9.3: Regional Shares of New Locations and Their Changes of Adjusted Shipments-Value and New Locations (All Manufactures, 1976-84)

Region		Regional Shares of New Locations		Changes	
		Total during 1976-81 (a)	Total during 1982-84 (b)	Value of Shipments (1982/76)	New Locations (b/a)
Hokkaido		6.91	4.76	-	-
Tokohu	N.	5.29	6.10	-	+
	S.	11.44	12.58	+	+
Kanto	I.	14.02	16.55	+	+
	C.	10.53	8.84	-	-
Tokai		11.93	11.04	+	-
Hokiriku		4.19	6.05	-	+
Kinki	I.	3.35	3.18	+	-
	S.	6.77	6.43	-	-
San'in		1.69	1.64	-	-
San'yo		5.88	5.15	-	-
Shikoku		4.94	4.53	-	-
Kyushu	N.	7.80	6.66	-	-
	S.	5.26	6.46	-	-
Japan	Total (%)	100.00	100.00		
	Number	10,306	6,095		

Note:
(1) 'New Locations' signifies merely the above mentioned sites acquired by firms. Usually most firms will start to construct plants there in a few years, or perhaps immediately as in recent cases of IC makers.
(2) For the changes of adjusted shipments value, see the left half on Table 9.2.
(3) + = Increase, - = Decrease in changes.

Source: MITI, Factory Location Trend Survey, surveyed annually through prefectural governments on the site (including a land expecting to be reclaimed) with area of more than 1,000m² acquired (including lease) by a firm to operate industry; unpublished.

better is its performance. This is by and large true of the middle distant regions (S. Tohoku, Hokuriku, San'in and Shikoku). These adjacent and middle-distant areas do not suffer such restrictions or such high costs of labour and land. (3) Of the most distant, North Tohoku and South Kyushu show increases and Hokkaido and North Kyushu decreases. Generally speaking these could be considered as an extension of the East-West reversal in economic status.

(b) **New Sites of Selected High Technology Industries**. We can see from Table 9.4 the total new sites of the selected seven high technology industries (three-digits at the classification level) between 1976 and 1981 and between 1982 and 1984.

Table 9.4: Regional Shares of New Locations and their Changes (Selected High Technology Industries, 1976-84)

Region		Total during 1976-81 (a)	Total during 1982-84 (b)	Changes (b/a)
Hokkaido		1.35	2.00	+
Tohoku	N.	9.48	11.48	+
	S.	21.20	21.31	+
Kanto	I.	25.62	23.85	−
	C.	12.93	8.77	−
Tokai		6.77	6.33	−
Hokuriku		3.83	4.99	+
Kinkai	I.	2.71	2.55	−
	C.	3.08	2.22	−
San'in		0.99	0.89	−
San'yo		4.19	2.55	−
Shikoku		1.72	2.55	+
Kyushu	N.	3.57	4.11	+
	S.	3.08	6.44	+
Japan	Total	100.00	100.00	
	Number	812	901	

Note: (1) + = Increase, − = Decrease.
(2) High technology industries: Medicines and drugs, telecommunications apparatus and allied products, electronic appliances, electrical measuring instruments, electronic apparatus parts, medical equipment and optical instruments and lenses.

Source: Factory Location Trend Survey, unpublished.

All of the established centres and their neighbours show a decrease in their share of new sites, whereas the middle-distant ones, except San'in, and all of the most distant (especially South Kyushu) show an increase. The new locations of high technology industries are expected to be more dispersed then those of all manufactures. From the East-West viewpoint, however, the concentration in the East, particularly the belt from Inland Kanto to North Tohoku through South Tohoku, is prominent and more intensive in the high technology industries than in all manufactures. The major reasons for this are the existence of the large industrial, especially machinery agglomeration, including top-level decision making and R&D functions (in Tokyo-Yokohama); the large intra- and international aviation centres (Haneda and Narita); the largest science city, (Tsukuba); containing 45 national and several private laboratories, cheaper labour and industrial sites; the openings of Skinkansen (in 1983); highways connecting with the Tokyo area (in Tohoku), and the presence of central government offices and many universities in Tokyo and the largest plain (Kanto Plain).

(c) **Conclusions**. We can draw three main conclusions. First, the dispersion of shipments-value by general sites has been proceeding to non-central areas. Second, this tendency is more noticeable in new locations (especially of high technology industry) than in shipment-values. Third, East Japan surpasses West Japan in order of prominence in (i) new sites of high technology industry, (ii) the sum of all industries and (iii) shipments-value.

4. **Review and Prospects**
The ability of high technology industry to move into new local areas cannot be denied, though there are opposing opinions. These locational changes seem to have arisen because of the increasingly improved transport facilities such as roads, highways, Shinkansens and jet-flight airports. It must be recalled that Weber (1909, 1914) in his theory of labour-cost orientation indicated the favourable effect on the labour orientation of industrial location through expanding a critical isodapane.

There are, however, some points to be kept in mind. First of all, older sites, the number of which must be far larger than that of new locations, can be employed for high technology activities by any firm or industry. When high-technology business is difficult in its present site or more profitable elsewhere, the firm or industry will (1) look for a new location, (2) consign the production to other plant(s) within the firm or to other companies including subsidiaries and/or subcontracted ones (as is often the case in Japan) and/or (3) acquire or merge with other firms (as is more usual in America). When the second option is adopted, this is not reflected in the number of new locations.

Central areas will probably undertake higher level functions when others operate high-technology industry. This is an aspect of the 'wave-ring' (or water ring) phenomena (Nishioka, 1983), meaning that new waves of activity arise at first or most powerfully, and successfully, in a central area and sooner or later, extend over more and more peripheral areas.

Of course, different products need different locational requisites for production. For instance, high quality labour is required for word-processing and industrial robots, abundant labour is necessary for VTR, facsimile, medicine and fine chemical ceramics, while both types are needed by IC, computer, medical electronics and mechatronics. For another example, pure water is preferred by IC and medicine, whereas abundant water is important for ceramics. It is therefore not strange to see high technology factories being situated even in relatively remote areas.

And yet, more generally, the companies whose products are easily transported and who attach importance to employing cheaper labour, whether it is a high technology or not, have jumped out from the original to newer, separate or distant plants; they have established themselves in suburban locations and have advanced to less developed areas long ago. The more the product is standardised in design, style, technology, process, delivery etc., the clearer the tendency will be. The original, established or traditional industrial areas and parent firms or plants would shift from their previous activities to those of higher value-added ones such as the production or R&D of non-standardised, fashionable, high quality, experimental, technologically most advanced and/or highly assembled product. Alternatively, they might take on new functions such as providing credit, technology, outlets, information material or components etc. to other companies or plants (Nishioka, 1976, pp. 214-

15).

Generally, the activity, which (1) needs a large number of physical/non-physical inputs, (2) produces a large number of physical/non-physical outputs, (3) places a vital importance on speedy and timely procurement, delivery and communication, (4) has to keep a close contact with its market or customers, suppliers and co-operators including subcontractors and nowadays system/software houses and (5) generates higher value-added per area of site, will tend to remain at its present site (if it is situated in an urban/industrial centre or area, has room for expansion and is not subject to any restriction) or possibly move to a nearby area or orientate itself to a large, well diversified, industrial agglomeration or metropolitan area (Nishioka, 1976, pp. 32-33). The concept of communication-oriented industry presented by Hoover and Vernon (1959, pp. 62-80) should be reconsidered. (See Figures 9.4 and 9.5, showing respectively relatively dispersed and concentrated distributions).

Figure 9.4: Main Factories of Semiconductor and Integrated Circuits (1983)

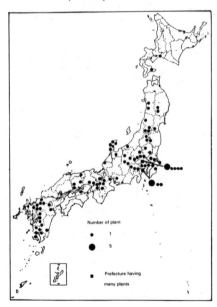

Source: JILC, 1985, <u>Industrial Location</u>, 24-7, pp. 54-56.

Figure 9.5: Main Factories of Telecommunications (1983)

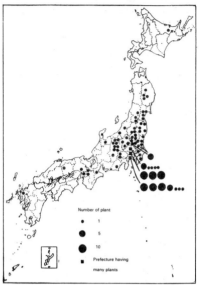

Source: JILC, 1985, <u>Industrial Location</u>, 24-6, pp. 56-59.

We should also recognise that different production stages can vary in their orientations. Integrated circuit (IC) production, for instance, is broadly divided into three stages: (1) R&D, (2) diffusing on wafer and (3) assembly and inspection. Tendencies suggest that (1) or (1) and (2) are operated in central areas, particularly in Kanto and (3) or (2) and (3) prevail in distant areas. In addition, ICs manufactured in distant areas are largely delivered to Tokyo metropolitan areas to be assembled into composite products (e.g. Shono, 1981; Hirano, 1981), or to be exported by air mostly through Narita. Such 'recurrent' phenomena (Miwa, 1983) - first metropolitan, then local and finally again metropolitan locations - can be widely seen in the machinery sector particularly the electrical machinery industry.

Also, leading corporations on many occasions set up divisional systems for their products and/or markets. Each division manages autonomously the manufacturing, selling, etc. to achieve the profit target as an organisational unit with an independent accounting system. In this case, R&D activities can be carried out in the following manner: (1) concentration in one particular division, (2) decentralisation to every division, or (3) co-existence of both of these. The first, which is called the central institute or laboratory, tends to be orientated towards a metropolitan area, to the company's head-office's or most important factory. The second is orientated towards the manufacturing plant or the market area. In the third case, the decentralised R&D might put emphasis on the product/process improvement (e.g. Dalbour, 1974). However, as a division of a subsidiary enterprise grows up, its R&D department will become larger or more specialised, and may deal with not only applied and development research but also specific basic research and can choose the location(s) away from the parent. Apart from this, the decentralisation of high technology industry containing R&D functions to distant areas would be, initially at least, occupied mainly in simpler parts assembly as well as the R&D of improvement types.

In short, not much effect, including technology transfer, on less-developed or depressed areas is expected, and many commentators are sceptical as to the effect of high technology industry on regional development.

At the same time, however, it must be said that there are some indications that less prosperous areas can benefit from the decentralisation of high technology industry. In fact, we have already seen the steady diffusion tendency of metal processing industry, especially from Coastal Kanto to the north-eastern half of Japan and even to South Kyushu (Silicon Island), gradually extending its sphere of influence. Changes at a corporate level, say Fujitsu, the largest computer manufacturer in Japan, have gradually expanded the geographical scope for locational arrangement of its establishments. Such products as small-sized computers may be developed by utilising VLSI towards fewer parts, more minimisation and mass production, which may make the location in local towns in further distant areas more possible. What the regionalist should attempt to do, at least, is to improve the circumstances which might otherwise constrain the above mentioned expansion of locational scope and if possible, try to encourage such expansions and, hopefully, establish another growth centre from which the locational scope will be expanded further and cover more distant areas.

There are still a large number of other factors to be considered concerning high technology industrial location and its distribution; such as corporate origins, historical backgrounds, top-management's philosophy or personality, and of course strategies, organisation, financial status, and particularly in Japan, labour relations.

For further study, it may be desirable, for example, to (1) classify more adequately products, stages, functions or activities, (2) divide labour costs into at least two and possibly, more categories, such as those for direct production, for operation programmes, for R&D etc, and (3) consider where (in terms of production stage, time and needless to say, place or area) and how much the value added arises.

Distribution of High Technology Industry in Tokyo Metropolitan Areas and Three Other Localities

Japanese high technology machinery industry, based mainly on microelectronics, is concentrated in the three largest metropolitan areas: Tokyo, Osaka and Nagoya. In particular, the Tokyo area with many head offices and R&D laboratories of big machine manufacturers, together with subsidiary and independent companies dealing in hardware and software, forms the largest technological centre because of this agglomeration. Osaka and Nagoya areas are also important but smaller technological centres. Some other local areas have much smaller high technology industrial agglomerations based upon the machinery industry. In the following section we shall observe the regional structures of high technology industrial agglomerations in the Tokyo area and some other local industrial areas.

The Tokyo Metropolitan Area. Kanto Region consists of seven prefectures and has the largest plain, most developed transportation systems and largest population (about 35 million in 1980) in Japan. High technology industry as well as population is concentrated particularly in the Tokyo area (i.e. southern Kanto) consisting of the four prefectures of Tokyo, Kanagawa (including the two big cities of Yokohama and Kawasaki), Chiba and Saitama (Figure 9.6). The share of southern Kanto, in terms of the number of plants in the whole of Japan, was approximately 20 per cent of integrated circuits, 50 per cent of industrial robot and medical equipment, 60 per cent of computer and space apparatus and 70 per cent of optical communication in 1982.

Figure 9.6: Factories of Selected High Technology
Industries (1982)

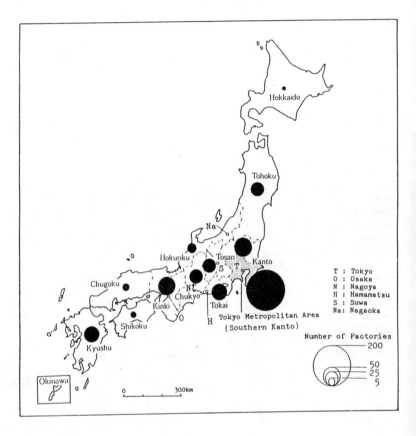

Note: High technology industries: IC, industrial
robot, medical equipment, optical electronic
appliance and aerospatial equipment.

Source: MITI, <u>Census of Manufactures</u>.

The relatively small number of IC plant is due to the mass production system and high bearing capacity of transport cost of IC, leading to the decentralisation of factories to such areas as Kyushu, Tohoku and Tosan.

The Tokyo area is the biggest industrial area in Japan, occupying 26 per cent of producer's shipments in 1980. For the machinery industry, its share in 1980 was about 30 per cent - almost the same as the percentage for the Osaka and Nagoya areas. The figures for the machinery industry are more than half in almost all divisions of the industry such as motorcars, electric machines, machine tools and precision machines, etc., and more than 70 per cent in many high technology products (Takeuchi, 1978, p.178).

Figure 9.2 shows the central part of southern Kanto where high technology industry is most concentrated. We shall divide this part into eastern, middle, southern and western Tokyo (in short, east, middle, south and west) and Kawasaki-Yokohama. Within this area, many high technology plants are located in a zone extending from the south to Kawasaki-Yokohama and its suburbs. Typical high technology plants such as Toshiba, Hitachi, SONY, NEC and Fujitsu in the outskirts of Kawasaki-Yokohama were originally generated in the south and have moved to their present sites since the 1960s (Takeuchi, 1973a, p. 105).

The middle area, which includes the Central Business District of Tokyo, has many enterprises' head-offices as well as central government offices, and forms the political and economic centre of Japan. It also forms the centre of wholesaling, other trading and also intra- and international information connected with all these functions. Publishing, newspaper, printing and their related industries are also concentrated there. (Takeuchi, 1980c, p.167).

The east area has been an industrial area for over 100 years. Its most important characteristic is in the agglomeration of factories producing various kinds of daily consumers' goods, particularly fashionable ones (Takeuchi, 1973b, p. 40). These factories are led by wholesalers situated in the middle area. Machines for producing such goods are also manufactured here, though their technological level is generally lower than that of the south area. Daily consumers' goods and related industries in Tokyo are now experiencing hard competition not only with those in other domestic areas, but also in

developing (and even developed, in the case of
fashion goods) countries. The industry in the east
area, which played the leading role until the 1950s,
is now stagnant and has become a significant element
of the inner city problems of Tokyo.

On the other hand, the south area, which is a
more recent industrial area than the east, is
characterised by the machinery industry and now high
technology industry. The machinery industry in the
south areas originally arose in an area adjacent to
the middle area, and until the second world war was
limited within the boundary of Tokyo prefecture.
After the war, especially since 1970, it has extended
to Kawasaki-Yokohama and to their suburbs. The
production system of the machinery industry is
characterised generally by a variety of assembling
functions. Motorcar plants such as Nissan and Isuzu,
electronics plants such as Toshiba, Hitachi, NEC,
SONY and JVC and large plants for machine tool and
watches, exist in outer areas of the Kawasaki-
Yokohama area or outskirts of the west. They function
as cores for the organisation a large number of
smaller factories of the machinery industry which are
concentrated mainly in inner areas such as the south
and nearby areas. The latter group contains almost
all of the machine and parts manufacturers, including
such fundamental processes as pressing, gilding,
processing, screw, spring, and gear manufacture, and
so on, showing as a whole a complicated
agglomeration. In the south, these small factories
are most densely distributed in Ota-ward (in the
south) at the eastern margin of which Haneda Airport
is located.

In southern Tokyo there are many subcontracting
machinery and other companies connected with one
another vertically, horizontally or diagonally,
forming an industrial-technological complex (Takeuchi, 1985, p. 325). The term 'industrial complex'
here signifies the situation in which various types
of industrial, particularly machinery, activities
have intimate and close relationships such as co-
operating or subcontracting. Large plants in the
outskirts have been supported by this industrial-
technical (later, rather technological) complex and
were originally developed in the technical centre of
the south (Takeuchi, 1980a, p. 159). Thus the south
played a 'seedbed' role for the machinery industry,
and still high technology firms are arising out of
this complex. The technological level of the parts
industry has been raised by the introduction of
microelectronic machines. High technology machinery

Figure 9.7: Distribution of Hi-Tech Factories and R&D
Labs in Central Tokyo Metropolitan Area (1982)

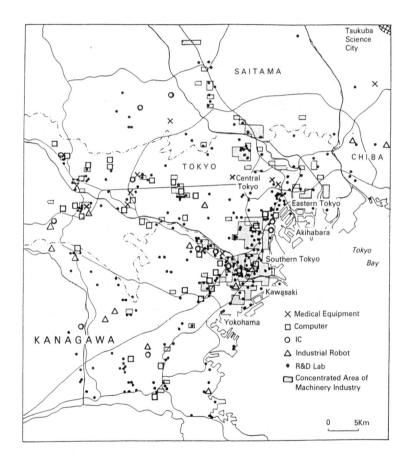

Source: MITI, <u>General Survey of Factories in Japan</u>,
1982. Science and Technology Agency, 1984.

plants in the Tokyo area also depend on this complex. In Kawasaki-Yokohama and the outskirts of the west, industrial parks for high technology industry, have been constructed by the local governments.

R&D has an important role in the functional system of high technology industry. Most national or private R&D laboratories are concentrated in the Tokyo area, including Tsukuba Science City. Tsukuba has been constructed as a national project and consists of a national university and national and public corporations' research facilities which were once in Tokyo Prefecture. In the Tokyo area there are most of the national research facilities and 90 per cent (1982) of their researchers. As to private research facilities, about 60 per cent of independent institutes and about 50 per cent of attached ones are in the Tokyo area, playing an important role in developing new products. They are concentrated in the middle and Kawasaki-Yokohama areas (Figure 9.7) and have a close contact with their head-offices. About 30 per cent (1981) of the university faculties or colleges of technology are concentrated in the Tokyo area, supporting R&D activities.

In the middle area is found a concentration of many companies involved in the development of products, including system houses, based on applications technology and microprocessor technology generally. In 1985 about 45 per cent of system houses were concentrated within Tokyo Prefecture, and half of them are in the Akihabara area. There are also concentrations in Minato-ward, which is also in the middle area, Ota-ward in the south, and two urban-subcentres, Shinjuku and Shibuya (in the middle area) (Figure 9.8). Akihabara has been famous as a concentrated district of wholesalers of electric household appliances selling at reduced prices. Lately, another function has been added there and almost all electronic products are now sold to the general public as well as professionals. Many people now visit there to buy products or to inspect pricing trends and to inspect new products. The agglomerated system houses and the nearby location of the University of Tokyo are inter-related. Thus, Akihabara is now characterised as the most interesting centre of technological information and development. System houses have connections with large companies' head-offices such as Toshiba, NEC, for example, which are located in Minato-ward, and also with small-sized machine-manufacturers which agglomerate in Ota-ward and which possess a reasonable level of technology.

Figure 9.8: Distribution of System and Software Houses in Tokyo Wards (1985)

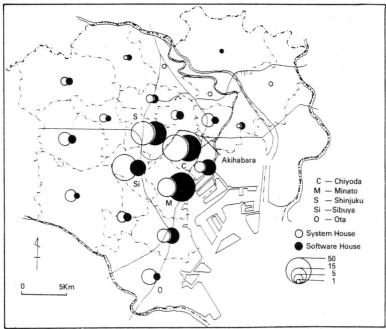

Source: Association of Information Industry, 1985a, 1986.

It can be said that without the assistance of system houses, Japanese high technology industry could not have developed so well. In fact, the large companies have relations with many system houses; for instance, NEC has links with 80 and Toshiba with 40 houses. On the other hand, software houses developing software for consumers' goods are concentrated in Minato-ward and two subcentres. Various enterprises in Minato-ward provide software houses with considerable business, and Shinjuku and Sibuya, which connect the metropolitan centre with residential suburbs, are advantageous locations. 80 per cent of all the information programming researchers and 60 per cent of all the system-engineers and programmers in Japan lived in Tokyo Prefecture in 1980.

These various reasons are why Tokyo is called by Takeuchi (1980b, p. 19) an organic 'complex' of R&D laboratories and development-type companies based upon a large number of machinery and metal-fabricating factories. This complex of high

technology industry is much stronger than those of
Osaka and Nagoya. Though today high technology plants
are distributed in every part of Japan, they are
generally under the strong influence of the Tokyo
area because of their far weaker R&D functions and
much lower level of technology. The actual conditions
of Japanese high technology industry should be
understood through considering this complex in the
Tokyo area and the nationwide systems or connections
based on the complex. Recently the high technology
industry of the Tokyo area has been spreading into
northern Kanto.

Some Local Areas. High technology industry has
recently been developing all over Japan and is
becoming one of the major characteristics of the
Japanese economy. The production of Integrated
Circuits in particular is distributed in many areas.
However, even in Kyushu, where a number of large
Integrated Circuit plants such as NEC, Toshiba,
Mitsubishi, Fujitsu, Sony, Oki, TI and so on are
located, and which is called 'Silicon Island' and
ranks first in the Integrated Circuit output in
Japan, there has so far been few R&D functions and
high technology subcontractors.

On the other hand, Hamamatsu in Shizouka
Prefecture, Suwa in Nagano Prefecture and Nagaoka in
Niigata Prefecture are the typical areas in which an
industrial 'complex' can be found (Figure 9.6). These
areas have such common features as (1) pre-existing
industries, (2) the existence of manufactures with an
artisan and self-help spirit, as well as an eagerness
for technological innovation and (3) relatively easy
access to the Tokyo area.

(a) **Hamamatsu area**. This area is connected with the
three largest metropolitan areas by Shinkansen and
high-ways. Based upon such antecedent industries as
textiles and wood-working machines, musical
instruments (Yamaha and Kawai), machine tool
manufacturing started. Various subcontractors have
appeared as these new industries have developed.
Production of motor cycles and/or motorcars by Honda,
Yamaha and Suzuki began on the basis of the
technological complex of the machine industry.
Hamamatsu is also producing electronic goods in
connection with colleges of industrial technology.
In the 1970s high technology industries such as
Integrated Circuits and computer and robots were

established. Hamamatsu is the most vital industrial city outside of the largest metropolitan areas.

(b) **Suwa area**. Suwa Basin, which is located in the mountainous area of central Japan, has made a remarkable progress in the electrical and precision machine industry. Before World War II, silk manufacturing was developed in the area, especially in Okaya city. In order to repair reeling machines, parts manufacturing such as valves, developed. During the war silk reeling was suspended and instead, the machinery industry was transplanted from Tokyo. After the war, based upon these plants, manufacturing of cameras, microscopes, music boxes and watches has developed, and now Suwa is the second centre of the precision machine industry in Japan next to the Tokyo area (Takeuchi, 1973a, p. 113).

After the war, Seiko drew back branch plants from local areas, but left only the Suwa Plant as it was superior to the Tokyo area in the level of technology. Suwa Plant has been leading the Seiko group in technological aspects and producing watches of high quality. Seiko has developed quartz watches since 1969 and has produced Integrated Circuit materials in the Suwa area. Related to this Seiko also established the EPSON company in Matsumoto near to Suwa in order to produce printers. EPSON is now one of the largest manufacturers of printers in the world. Integrated Circuits, robots and various kinds of microelectronic machines and their parts factories are distributed in this area and Suwa has become one of the typical high technology industrial areas of Japan.

(c) **Nagaoka area**. Nagaoka which faces the Japan Sea, was once an oil-producing field and manufactured drilling machines. Based upon oil machine manufacturing, a machine tool industry was established. After the war, because motorcar and electronics industries which supported Japanese economic development had not yet been established there, the industrial activity of Nagaoka became stagnant (Takeuchi, 1983b, p. 43).

However, by inviting high technology plants as well as researchers and workers to make good use of the opening in Shinkansen (in 1982) of new highways and a national college of technology, and of the designation of the area as a Technopolis (in 1984). Nagaoka has been making a great effort to construct a

new town of large scale industrial parks. There is
now a problem of getting the induced college and
plants to co-operate well with the existing local
industrial complex.

Conclusion

From the above, it is clear that new high technology
industrial areas in Japan have been strictly limited
in their number, and have developed largely by a
long-run and successive internal effort of
endogeneous progress made by indigenous (existing or
newcoming) industries. In these areas there are still
few R&D laboratories and development-type factories
and a weak basic industrial stratum. So far, these
areas have generally worked as a subsystem of the
national network, organised and led by major
metropolitan industrial areas. It is, of course, a
very difficult but important task for both industries
and communities in local areas to raise the
technological level. For this purpose, it is
necessary to strengthen R&D functions and to
encourage development-type firms including software
and system houses.

Appendix A
The Technopolis Concept and Related Arguments

Probably backed by the success of Kyushu, which
contributed about 40 per cent of IC production in
1978, and expecting high technology industry to play
a vital role to revitalise rural or stagnated areas
in the future, MITI advocated in 1980 the
'Technopolis Concept'. This concept suggested the
creation of cities in which 'industry' (advanced
technology industries including electronics and
machinery), 'science' (engineering departments of
universities or technology colleges and research
institutes or laboratories) and 'comfortable life
environment' (residences and social and cultural
facilities or services) are brought together in an
organic relationship, and where traditions and
nature are merged in a modern industrial structure.
In 1983, MITI enacted the Law for Accelerating the
Regional Development Based Upon High-Technology
Industrial Complexes (in short, the Technopolis Law)
to help promote the construction of advanced
technology-concentrated areas, and began to
designate the Technopolis Region in March, 1984 the
time of writing this paper, fifteen, including
Hamamatsu and Nagaoka, of all nineteen candidate

regions have been designated.

Different views have arisen on the prospects of the Technopolis concept. The optimistic view, backed by the great success of Silicon Island (Kyushu, the third and western-most situated of Japan's four largest islands), insists that high technology products exhibit so high a bearing capacity of transport costs, owing to very high value-added and price per unit weight, that they can easily be located in remote local areas which are connected through high speed transportation facilities to central areas. Moreover, the optimists suggest that the future developments of new information and communicating systems will diminish the communication-cost differentials derived from distance differences, so that the dispersion of high technology industry into local areas will necessarily become much easier.

On the other hand, the pessimists or sceptics argue that high technology industry generally must orient towards the largest metropolitan areas since it needs researchers, engineers, highly skilled workers, a variety of related industries and easy access to the latest information. Some of them emphasise the promotion of 'Jiba Sangyo' (3) rather than the attraction of large manufacturers from central areas.

Both opinions may be partly correct and partly incorrect. The optimistic view on Technopolis results from the confusion of distantly locatable products or production stages with those characterised by metropolitan or agglomerative orientation. Also, though the optimists explicitly or implicitly maintain that if new information and communication systems develop in the near future, everyone can get any information, anywhere and anytime, at the almost same cost, they overlook that there are important types of information, which will not necessarily have these features, the presence or absence of which can distinguish some areas from others. On the other hand, the sceptics' view neglects the considerable possibility of dispersion of at least a part of high technology industry, and disregards the differences in nature between the Technolopolis movement to be initiated by localities and previous regional developments led mainly by the central government.

Both, however, could be integrated because, they give consideration to the technological progress in Jiba Sangyo, albeit from different viewpoints. The optimists stress the fact that today's high technology is so suitable for multifarious production in small quantities that small firms could and should improve their technological, and thus competitive status through adopting or applying high technology products. Some pessimists have attached much greater importance to small firms in localities, which use local materials, employ local people, develop local markets, are operated by local entrepreneurs and financed by local capital, and hence, can contribute to sustain the regional economy.

Although when the Technopolis Concept was first devised, the relation between Jiba Sangyo or local small firms and high technology might have been considered to be meagre, the final Concept clearly has recognised at least two different courses. One is led mainly by the attraction of high technology industry from central areas, which is expected to transfer high technology to local small firms, and which is expected to provide an attractive technological base for such central firms. Secondly, unlike the pessimist's supposition, the Technopolis Concept in fact never intended to disperse high technology industry to remote areas by coercion at all. Rather it stresses the existence of a 'mother city' within or nearby a proposed Technopolis Region, which has about 150,000 or more population and provides urban and cultural facilities or services to newcomers working for incoming high technology activities, as one of the desirable requisites for a would-be or proposed Technopolis.

Therefore, the Technopolis Concept can be considered to have presented a way of integrating both the different views on high technology industry and regional development.

One of the most significant impacts of the Technopolis Concept on regional development has been to arouse public opinion to new possibilities, necessities and to measures for regional development, reconstruction or redevelopment in many areas whether they are Technopolises or not.

Notes

1. According to recent interviews by Nishioka with managers in San'yo Region in May and June, 1985, seemingly old-fashioned establishments have made various efforts to innovate processes and/or products.

Kurray in Kurashiki, Okayama Prefecture for instance, which has produced rayon since before the

Appendix B
Districts (most conventional) and Prefectures of Japan

PREFECTURES

1 Hokkaidō	25 Shiga
2 Aomori	26 Kyoto
3 Iwate	27 Osaka
4 Miyagi	28 Hyōgo
5 Akita	29 Nara
6 Yamagata	30 Wakayama
7 Fukushima	31 Tottori
8 Ibaraki	32 Shimane
9 Tochigi	33 Okayama
10 Gunma	34 Hiroshima
11 Saitama	35 Yamaguchi
12 Chiba	36 Tokushima
13 Tokyo	37 Kagawa
14 Kanagawa	38 Ehime
15 Niigata	39 Kōchi
16 Toyama	40 Fukuoka
17 Ishikawa	41 Saga
18 Fukui	42 Nagasaki
19 Yamanashi	43 Kumamoto
20 Nagano	44 Oita
21 Gifu	45 Miyazaki
22 Shizucka	46 Kagoshima
23 Aichi	47 Okinawa
24 Mie	

Note: Though the district dividing and prefectural-numbering ways hereof are most conventional and often adopted, there are many different ways for specific purposes.

war and vyniron and other petrochemical polymers
after the war, is developing chemicals, medicines and
new ceramics. Also, the Toyo Soda's Nanyo Plant in
Yamaguchi Prefecture, which was established in 1936
for producing soda through an ammonia process and
then electrolysis process, has proceeded to produce
since the war, cement and chemical fibre and
fertilisers and after the oil crisis, several
advanced goods such as high speed performance liquid
chromatography and related products.

The share of high tech products in the total
sales for the Nanyo Plant, according to its managers,
is at least 20 per cent and will be more than 45 per
cent in the future, whereas in case of Kurray, that
of textiles still occupies about 70 per cent and that
of high technology products, only two per cent or so.
The share of the newest products in the total sales
or profits for a plant or company belonging to a
matured industry seems to be between the above both
cases and presently on average, probably nearer to
the latter case.

2.   For a map of all 47 prefectures and the
eight most conventional regions, see Appendix B.

3.   Jiba Sangyo refers to each or a group of
small industries localised in an area, usually
operated in small establishments by local
entrepreneurs with locally generated capital, and
often organised by the Ton'ya or Toiya (Wholesaler)
or by the Seizo-Oroshi (manufacturing-wholesaling
contractor). Most of their products can be sold
nationally and internationally. Jiba Sangyo are
distributed throughout the country, with approximately one-fifth being situated in Kinki Region,
including Osaka and Kyoto and Kanto Region including
Tokyo. Compared to the population distribution,
these firms are disproportionately numerous in the
outlying regions. (For details, see Ide and Takeuchi,
1980, pp. 299-319).

References
(J) denotes that the original article is in Japanese.
Aoki, T. (J) (1960), 'On the Advantage of Industrial
Agglomeration: An Economic Geographical Study,'
Annals of Hitotsubashi University, Studies in
Economics, 4, 259-320
Association of Information Industry (J), (1985a),
Directory of Software Houses in Japan
Association of Information Industry (J) (1985b),
Directory of System Houses in Japan
Dalborg, H., (1974), Research and Development:

Organization and Location. Stockholm School of Economics
Hirano, M. (J), (1981), 'Technopolis and Industry,' Industrial Location (JILC), 20-7, 6-17
Hoover, E.M. and Vernon, R., (1959), Anatomy of a Metropolis Harvard University Press
Ide, S. and Takeuchi, A. (1980), 'Jiba Sangyo; Localized Industry,' ccc The Association of Japanese Geographers, (ed.), Geography of Japan (Teikoku-shoin, Tokyo), 299-319
Ito, K. (J), (1985), 'Trend of Industrial Location and How to Equip Coastal Estates,' Industrial Location, 24-2, 15-19
JILC (Japan Industrial Location Centre), (J), 1984, unpublished Data of Shipments-Value, based on the Census of Manufacturers with four or more persons engaged, adjusted by price indices and computed at the 1975 prices by JILC
JILC (J), (1985), 'Factory Distribution by Industry,' Industrial Location, 24-6, 48-61, 24-7, 54-63
Kawashima, T., (1964), 'Regional Structure of Japanese Industry,' Annals of Economic (Osaka City University), 21, 1-23
Kawashima, T., (1980), 'The Regional Pattern of Japanese Economy: Its Characteristics and Trends,' AJG, (ed.), Geography of Japan, 390-414
MITI (Ministry of International Trade and Industry), annually, Census of Manufacturers
MITI (J), annually, General Survey of Factories in Japan
MITI (J), annually, unpublished, Factory Location Trend Survey
MITI , (1984), 'Preliminary Report on the Census of Manufacturers (with four or more persons engaged). (1983),' Industrial Statistics Monthly, 37-11, (J), 4-17; Eng. transl. 18-21
Miwa, K. (J), 1983, 'Behaviour of Industrial Decentralisation under the Change of Industrial Structure in Japan,' Annals of the Japan Industrial Location Centre, (JILC), 10, 3-26
Nishioka, H., (1962), 'Interregional Economic Differences within Japan,' Aoyawa Journal of Economics (A.G.U.), 13-4, 37-57
Nishioka, H. (J), (1963a), Location and Regional Economy (Miyai-shoten, Tokyo)
Nishioka, H. (J), (1963b), 'Interregional Economic Differentials and Location Policy in Japan,' Annual of Japan Economic Policy Association, 11, 86-96
Nishioka, H. (J), (1966), A Study on the Interregional Income Differentials in Japan (Kokon-

shoin, Tokyo)
Nishioka, H. (1967), 'On the Interregional Income Differentials in Japan,' Papers and Proceedings of the Second Far East Conference of the Regional Science Association, 1965, 169-80
Nishioka, H. (J), (1976), Economic Geographical Analysis Taimeido, Tokyo
Nishioka, H. (J), (1981), 'Japanese Regional Development and Industrial Location, and Some Macroscopic Understandings', The Human Geography, 33-6, 64-73
Nishioka, H. (J) (1983), 'Location of New High Technology Industry', Aoyama Journal of Economics, 35-3, 66-88
Nishioka, H. (J), (1984), 'Advanced High Technology,' Annals of the Association of Economic Geographers, 30-4, 263-77
Nishioka, H. (1985), 'High Technology Industry: Location, Regional Development and International Trade Frictions,' Aoyama Journal of Economics, 36-2, 3 & 4 (combined), 295-337
Nishioka, H. and Oshiro, K.K., (1979), (1980), 'Industrial Location and Policy in Japan, 1945-1970,' I and II. Aoyama Journal of Economics, 31-1, 35-52; 31-4, 133-46
Science and Technology Agency (J), 1984, Directory of Laboratories in Japan
Takeuchi, A. (J), (1973a) Japanese Machinery Industry (Taimeido, Tokyo)
Takeuchi, A. (J) (1973b), Regional Structure of Concentrated Area of Manufacturing Industry in the Large City, Annals of the Association of Economic Geographers, 19-2, 40-57
Takeuchi, A. (J), (1978), Regional Structure of Manufacturing Industry, Taimeido (Tokyo)
Takeuchi, A. (1980a), 'Motor Vehicles,' Murata, K. (ed.) An Industrial Geography of Japan Bell Hyman, London, 152-62
Takeuchi, A. (1980b), 'The Industrial System of Tokyo Metropolitan Area,' Report of Researches (N.I.T.), 7-2, 1-40
Takeuchi, A. (J), (1980c), 'Regional Structure of the Printing Industry in Tokyo,' Geographical Reports of Tokyo Metropolitan University, 14 & 15, 163-73
Takeuchi, A. (J), (1983a), Technical Complex and Industrial Community, Taimeido, Tokyo
Takeuchi, A. (J), (1983b), 'The Technical Complex of Machinery Industry in Nagaoka City,' Annals of the Association of Economic Geographers, 29-2, 107-19
Takeuchi, A. (J), (1985), 'The Complex Area of Machinery Industry in Tokyo Metropolitan Area.'

Report of Researches, 14, 317-27
Weber, A. (1909), Ueber den Standort der Industrien,
I (Mohr, Theubingen). Eng. trans. by Friedrich, C.J.,
1929, Theory of the Location of Industries
(University of Chicago Press)
Weber, A. (1914), 'Industrielle Standortslehre,'
Grundriss der Sozialoekonomik, 6 Abt. (Mohr,
Tuebingen), 58-86

Chapter Ten

H.T.U.K. THE DEVELOPMENT OF THE UNITED KINGDOM'S MAJOR CENTRE OF HIGH TECHNOLOGY INDUSTRY

Michael J. Breheny and Ronald McQuaid

Introduction

If in the course of history a single phrase becomes associated with the 1980s it is likely to be 'high technology', or more colloquially 'high tech'. In just a few years this label has been attached to all manner of activities and artefacts; from advanced industrial processes, to modern buildings, to running shoes, to furniture, to methods of childbirth, and many more. This is no less the case in the United Kingdom than elsewhere.

The phrase is used in many cases in a trivial sense, but when applied to industrial activity it takes on a profound significance. As older manufacturing industrial sectors decline, the hope in many countries is that growth in high technology industries will compensate. Consequently these newer industries are given varying degrees of privileged status, both by governments and public opinion. Those localities that are claimed to have a concentration of high tech industries are envied by other less fortunate areas, and local authority industrial initiatives often aim to replicate those local factors tht are assumed to attract high tech industrialists.

This optimistic interest in 'high tech' is perpetuated by the 'hype' of both a media interested in any modern activity and a property profession anxious to create a new high tech industrial market. However, in the U.K. at least, this optimism has not been founded on any serious assessment of the merits of the popular claims for high tech industries. Simple questions such as: how much output and how many jobs have high tech sectors created?; what kind of jobs are created?; why does high tech industry locate in some areas and not in others? what contribution is high tech likely to make to our future national and local economies? This paper is based on a study which attempted to provide answers to these and related questions. The study, which was carried out during 1984 and 1985 in the Department of Geography, University of Reading, England, is reported more fully in Hall et al, (1987).

The aim of the study was to assess the genesis and development of high tech industry in an area – the so-called 'M4 Corridor' from London westwards to Bristol and South Wales – which was popularly assumed to be Britain's answer to Silicon Valley, California. This paper reports some of the findings of this study, including the adoption of a preferred locational label: the 'Western Crescent'.

Although the ultimate focus of the paper is on this particular area, the context is set by a review of national high tech industrial changes. Hopefully, this combination of broader assessments and considerations of the U.K.'s most important case study location will go some way to addressing the important questions raised above.

The second section of the paper begins the contextual assessment by looking at recent national and regional changes in employment generally. For the purpose of commencing the investigation, the paper adopts at this stage the popular notion of a distinct 'M4 Corridor' and assesses overall employment change in those counties involved. Before moving on from employment generally to 'high technology' employment in particular, it is obviously necessary to define what is meant by the term. This is done in section 3. Section 4 begins the locational analysis by drawing up a national geography of high tech jobs.

These sections are followed in section 5 by an analysis of the reasons behind the growth of high technology industries in what by that stage has been identified as the 'Western Crescent'. This section looks initially at the 'spatial divisions of high technology labour', and goes on to assess the early development of the area. As part of the attempted explanation, section 5 also discusses the results of a survey of high technology firms in Berkshire, carried out as part of the original project. The section finally considers the importance of the role of defence spending on the development of the 'Crescent' and the possible impact of current changes in defence procurement policy. This is followed in section 7, by concluding comments.

Table 10.1: Employment Change, South East England, 1971-81

	Manufacturing		Services		Total	
	Abs.	%	Abs.	%	Abs.	%
Inner London	-186,722	(-40.6)	-105,041	(-6.1)	-318,300	(-13.9)
Outer London	-192,047	(-32.6)	98,681	(10.2)	-95,662	(-5.8)
OMA	-125,370	(-18.4)	237,233	(25.0)	111,604	(6.3)
OSE	-15,337	(-3.5)	231,111	(23.9)	210,487	(13.6)
South East	-519,476	(-23.9)	461,984	(10.0)	-91,871	(-1.3)
GB	-1,963,200	(-24.9)	1,703,000	(14.5)	-501,200	(-2.3)
Berks, Bucks, Hants,						
West Sussex	-11,106	(-3.0)	208,190	(32.1)	196,595	(17.6)
Other ROSE counties	-129,601	(-17.2)	260,154	(20.5)	125,496	(5.7)

Source: SERPLAN (1984) (employees in employment, 1968 SIC).

Recent National and Regional Changes in Employment

In order to put recent changes in the 'Corridor' into perspective we need to see to what extent these changes differ from those taking place in a broader space-economy. It is useful, then to look at changes in employment both nationally and regionally in recent years.

The general picture nationally is of declining levels of employment and dramatically increasing levels of unemployment. This general picture disguises decline in the manufacturing sector (1,963,200 jobs, or 24.9%), lost from 1971 to 1981 in Great Britain and modest but significant gains in the service sectors (461,984 jobs, or 10.0% gained 1971-81). These relative changes between manufacturing and service sectors reflect a trend in all industrialised nations, but the decline in manufacturing has been more marked in Britain than in most other such countries. As will be shown later, the high technology manufacturing sector also contributed to the loss of jobs.

A useful general picture to have in mind when considering localised change in Britain is of two related and overlapping trends in the space-economy. One is that the south of England continues to prosper relative to other regions, and that its relative advantage is probably being reinforced at the moment partly because of its success in the service and high technology sectors. The second trend is that of general movements in economic activity away from urban areas out towards suburban or largely rural areas.

Table 10.1 reflects both of these trends. On the face of it, the South-East in general has not done particularly well relative to Great Britain as a whole. Its manufacturing performance (-23.9% 1971-81) is only marginally better than the nation (-24.9%), and its service sector employment (+10.00%) is poorer than that for Great Britain (+14.5%). Overall, it fared only marginally better than Great Britain (-1.3% compared to -2.3%). However Table 10.1 makes it clear that the South-East's overall performance is heavily weighted by London's, and particularly Inner London's, performance. Performance in the Outer Metropolitan Area and the Outer South-East are considerably better on all counts than Great Britain or the South-East as a whole.

Table 10.1 demonstrates very clearly the trend in the movement of economic activity away from major cities. Apart from the performance of the Outer Metropolitan Area (OMA) in service employment, the

Table 10.2: Employment Change in non-South East 'Corridor' Counties, 1971-81

	Manufacturing		Services		Total	
	Abs.	%	Abs.	%	Abs.	%
Wiltshire	-13,956	(-21.3)	29,804	(32.7)	16,442	(9.4)
Avon	-25,264	(-20.1)	29,914	(13.6)	2,798	(0.7)
Gwent	-19,763	(-27.5)	8,869	(12.4)	-11,767	(-7.3)
South Glam.	-16,896	(-39.8)	3,531	(2.9)	-16,683	(-9.3)
Mid Glam.	-10,656	(-16.3)	12,227	(14.6)	-8,331	(-4.9)
West Glam.	-22,872	(-36.5)	21,977	(32.5)	-9,948	(-6.5)

Source: Annual Census of Employment (employees in employment, 1968 SIC; manufacturing = SIC's 3-19, services = SIC's 21-27).

percentage change figures show a neat and consistent pattern of high levels of decline nearer the urban core and increasingly better performance away from it.

The point of stressing the two trends of southern advantage and suburban/rural advantage is that with these conditions prevailing nationally, we would expect areas such as the 'Corridor', to be performing relatively well regardless of concentrations of high technology industry.

The evidence presented here is entirely consistent with these two trends. However, the 'rings' of relative advantage around London are not homogeneous. They do show consistent advantage away from London, but with a distinct western bias. As Table 10.1 shows, the four western Rest of South East (R.O.S.E.) counties have together performed rather better than the R.O.S.E. as a whole. A decline of only 3.0% in manufacturing, a rise of 32.1% in services and an overall growth of 17.6% over the 1971-81 period is a considerably better performance than that of other R.O.S.E. counties. Indeed, the four counties took over half of the total R.O.S.E. employment growth during the period.

If we look at 'corridor' counties not in the South-East region (i.e. those further west), we find that the pattern of manufacturing decline and service growth is repeated consistently. Table 10.2 shows employment change in the two broad sectors, plus overall change, for six counties. Wiltshire has the best overall performance with service sector growth of nearly 33%. Change in the two sectors in Avon over the 1971-81 period almost balance out.

The four South Wales counties all show overall employment losses, but from very different rates of change in the two sectors. South Glamorgan has fared worst because of a high manufacturing loss and a low service gain. The most dramatic restructuring has taken place in West Glamorgan, where a 36.5% decline in manufacturing has been countered by a 32.5% increase in the service sector.

This simple analysis of recent employment change nationally, regionally and in particular 'Corridor' counties provides a necessary context for looking at the contribution to these changes of high technology industry. These general employment changes provide us with an initial idea of why this high technology industry has grown in particular locations. Possibly more importantly, the magnitude of these general changes cautions against any undue excitement about the prospect of this new growth making major inroads into our unemployment problems.

Defining 'High Technology Industry'

A major problem for all researchers investigating so-
called 'high-technology' industry is simply that of
defining it. As yet, as many definitions have emerged
as there are research projects (see for example,
McQuaid and Langridge 1984, Ellin and Gillespie 1983,
SERPLAN 1983, Longhurst, 1983, Markusen, Hall and
Glasmeier, 1986, Hall et al, 1987).

The reason why definition is so problematical is
that researchers are trying to compromise between
devising a conceptually sound, consistent and
exhaustive definition and one which allows
measurement and is practicable. McQuaid and
Langridge (1984) seek to define a 'core' set of 'high
technology' industries on the basis of their
occupational structures (an above average share of
engineers, technologists and scientists) and the
relative amount that they spend on research and
development. While there are obvious problems of
heterogeneity within Minimum List Heading industries
they identify seven 'core' high-technology
manufacturing industries (Table 10.3). These
industries are aerospace, electronic computers,
radio, radar and electronic capital goods, radio and
electronic components, telegraph and telephone
apparatus, broadcast receiving equipment, and
pharmaceuticals, on the 1968 Standard Industrial
Classification (SIC) classification of industries. A
1980 S.I.C. version of this is also shown in Table
10.3. Whilst this latter classification gives a more
up-to-date set of industrial groupings, it is of
little use in looking at time series data. For this
reason, most of the analysis presented here is of
high technology industry defined on the 1968 basis.

McQuaid and Langridge's (1984) definition,
which is adopted here, concentrates upon the
producers of high technology goods rather than the
users, although many of the greatest impacts of
technology are in the latter industries (see Braun
and Senker, 1982). The producer industries are
concerned with new, high technology products (such as
semiconductors, computers, robots etc.) and the user
industries concerned with the use of these products
in improving production processes, often in
traditional manufacturing sectors (such as the use of
robots in car manufacture). Whilst acknowledging the
importance of both product and process innovations to
sectoral, national and local economic development,

Table 10.3: High technology industries defined by industrial sector on 1968 and 1980 classifications

1968 SIC

MLH 272	Pharmaceutical chemicals and preparations
MLH 363	Telegraph and telephone apparatus and equipment
MLH 364	Radio and electronic components
MLH 365	Broadcast receiving and some reproducing equipment
MLH 366	Electronic computers
MLH 367	Radio, radar and electronic capital goods
MLH 383	Aerospace equipment manufacture and repair

1980 SIC

AH 2570	Pharmaceutical products
AH 3302	Electronic data processing equipment
AH 3441	Telegraph and telephone apparatus and equipment
AH 3442	Electrical instruments and control systems
AH 3443	Radio and electronic capital goods
AH 3444	Components other than active ones, mainly for electronic equipment
AH 3453	Active components and electronic sub-assembly
AH 3454	Electronic consumer goods and other electronic equipment
AH 3640	Aerospace equipment manufacture and repairing
AH 7902	Telecommunications (excluding Post Office, broadcasting and local cable relay systems)

the focus of our attention, and the basis of our definition, is with high technology producers.

If the popular assumption - that high technology industries are amongst the few exceptions to general manufacturing decline - holds, we might expect to find our seven M.L.H.'s performing well at the national level. Sadly, this is not the case.

National employment changes in our high technology group of industries are shown in Table 10.4, for the 1971-1981 and 1975-1981 periods for the 1968 based definition, and, in Table 10.5, for the 1980-84 period for the 1980 based definition. Use of data for the 1975-81 period is preferable, because it controls for the business cycle (both years were 'troughs'). It also reflects the recent activity that we associate with the growth of high technology

Table 10.4: Employment Change in aggregate High Technology Sectors, 1971-83, on 1968 SIC base

MLH	June 1971	June 1975	Sept. 1981	Sept. 1983	Change 71-81	Change 75-83
272 Pharmactl Chems, Preps	72,600	76,100	69,300	71,800	-3,300	-4,300
363 Telegraph, Telephone appts, equmt	84,300	87,000	64,500	56,200	-19,800	-30,800
364 Radio, electronic components	128,300	128,400	103,100	106,800	-25,200	-21,600
365 Broadcast receiving equipment	48,300	54,900	37,500	22,500	-10,800	-32,400
366 Electronic computers	50,100	43,300	42,100	58,100	-8,000	14,800
367 Radio, radar, electronic capital goods	94,100	89,300	100,100	105,300	6,000	16,000
383 Aerospace equipment mfr, repair	211,400	204,400	197,300	163,000	-14,100	-41,400
Total	689,100	683,400	613,900	583,700	-75,200	-99,700

Sources: British Labour Statistics; Department of Employment Gazette

industry in recent years. Use of recent change data should go some way towards separating out activity in the new high technology firms in each industrial sector.

Table 10.5: Employment change in aggregate High Technology Sectors 1981-4, on 1980 SIC Base. NB 3 digit figures used where 4 digit figures not available.

Activity Heading	Dec. 1981	Sept. 1984	Change 81-84
2570 Pharmaceutical products	81,200	82,300	1,100
(330) Electronic data processing equipment	72,200	74,700	2,500
(344) Radio, electronic capital goods, components	200,500	205,100	4,600
(345) Electrical, electronic consumer goods, components	125,300	135,700	10,400
3640 Aerospace equipment manufacture, repairing	182,100	157,800	-24,300
7902 Telecommunic-ations	231,700	224,200	-7,500
Total	893,000	879,800	-13,200

Source: Department of Employment Gazette.

We can see immediately from Tables 10.4 and 10.5 that at the national level the much vaunted high technology group contributes not to employment gain, but to substantial loss. On the 1968 definition only two of the sectors, Electronic computers (M.L.H. 366) and Radio, radar and electronic capital goods (367) actually increased their employment levels over the 1975-83 period. On the more sensitive 1980 definition, we still see an overall loss of employment. Whilst the performance of these high tech manufacturing sectors is clearly better than manufacturing industry as a whole, it is very disappointing given the importance attached to these sectors.

But is this dismal performance of the high tech sectors just a British phenomenon; just a further reflection of the relative decline of the British economy? Unfortunately, it appears that it is - certainly if we make comparisons with the United

Table 10.6: Great Britain and the United States: High Tech Employment

Great Britain				United States		
	1981 000s	1971-1981 000s	%	1981 000s	1972-1981 000s	%
Pharmaceutical Preparations	69.3	-3.3	-4.6	130.5	+18.5	+16.5
Telephone, Telegraph App.	64.5	-19.8	-23.5	147.4	+13.0	+9.7
Radio, TV Transmission etc.	103.1	-25.2	-19.6	426.9	+107.7	+33.7
Electronic Computing Eqmt.	37.5	-10.8	-22.4	320.7	+175.9	+121.5
Aircraft	42.1	-8.0	-16.0	301.1	+69.3	+29.9
Electronic Components, NEC	100.1	+6.0	+6.4	190.0	+89.5	+89.1
Semiconductors, Related Devices	197.3	-14.1	-6.7	169.5	+71.9	+73.6
Construction Machinery Eqmt.				145.9	+12.1	+9.6
Aircraft Parts, etc.				140.3	+38.1	+37.3
Aircraft Engines, Parts				140.0	+35.3	+33.9
Total (10) / Total (7)	613.9	-75.2	-10.9	2112.3	+631.3	+29.9
Total (100) / Total (43)	3253.2	-366.2	-10.1	5475.2	+1080.9	+24.6

Great Britain categories (left): Pharmaceutical Chemicals etc.; Telegraph and Telephone; Radio and Electronic etc.; Broadcast Receiving etc.; Electronic Computers etc.; Radio, Radar, etc.; Capital Goods; Aerospace Equipment etc.

Source: Hall et al, (1987)

States. Hall et al (1987) have compared growth in the seven U.K. sectors with an equivalent set of high tech industries in the United States. Table 10.6 shows that whilst the UK group accounted for an overall loss of 75,200 jobs (-10.9%) over the 1971-81 period, the U.S. group contributed job growth of 631,300 jobs (+29.9%). In both cases, a comparison is also made with a 'long-list' of high tech industries. The 43 U.K. industries lost 366,600 (-10.1%) jobs, compared to a growth in the 100 U.S. industries of 1,080,700 jobs (+24.6%).

Clearly, then, the popular assumption about the positive role of high tech industries does hold in the United States. Over 1972-81 they were responsible for a massive direct job gain. By comparison, the U.K. performance was appalling.

The findings from this simple comparison are important, because they go some way to answering the question often posed, in the U.K. at least; can high technology industry be a major source of new jobs? The answer seems to be that it depends - on the vigour of the economy in question. In the U.S., and possibly Japan, it seems that the rate of growth in high tech can be sufficiently high - despite obvious productivity gains - to offset a large number of job losses in traditional manufacturing industries. In the U.K. there is no sign that the economy can be sufficiently vigorous for this to happen.

This argument is rather circular, of course, because it may be high tech industry that has created a vigorous U.S. economy rather than the other way around. Nevertheless, the American experience does seem to disprove the pessimistic view often expressed in Britain, that high tech industry is incapable, in principle, of creating large numbers of jobs.

Because the aim here is to go on to look at the geography of high technology industry in Britain, we are forced to use employment as our measure of the phenomenon. Other measures are just not available at local levels. However, whilst we are reviewing high tech at the national level we can make use of data on industrial output. Given the nature of high tech industry we might expect its disappointing job-creating performance in the U.K. to be due in part to exceptional gains in productivity. Table 10.7 demonstrates that, with the notable exception of the Electronic Computers M.L.H., our seven high tech sectors had neither exceptional growth in output over the 1971-78 period or in output per employee. All seven high tech sectors together had a growth in output of 15.02% and growth in output per head of

Table 10.7: Changes in Output and Productivity for U.K. High Tech. Industries, 1971-78 (1978 prices)

	1971 Output (£000's)	1978 Output (£000's)	% change Output	1971 Output per head (£'s)	1978 Output per head (£'s)	% change Output per head
MLH 272 Pharmaceuticals Chemicals and Preparations	824,454	1,008,806	22.36	13,383	15,147	13.2
MLH 363 Telegraph and Telephone Apparatus	490,285	447,005	-8.82	4,998	6,613	32.3
MLH 364 Radio and Electronic Components	694,173	708,213	2.02	5,522	6,443	16.7
MLH 365.2 Broadcast Receiving and Sound Equipment	336,767*	151,756	-54.94	7,961	4,775	-38.7
MLH 366 Electronic Computers	186,215	431,950	131.96	7,920	18,118	128.7
MLH 367 Radio, Radar and Electronic Capital Goods	557,682	869,393	55.89	6,530	8,383	28.4
MLH 383 Aerospace Equipment Manufacturing	1,174,932	1,288,149	9.63	5,485	6,709	22.3
All high tech sectors	4,264,586	4,905,272	15.02	6,553	8,238	25.7
All manufacturing industry	45,119,791	58,110,400	28.56	5,772	8,065	39.7

Sources: Census of Production, 1975 (Business Statistics Office) Census of Production, 1978 (Business Statistics Office)

25.7% compared to figures for all manufacturing industry of 28.56%, and 39.7%. Only MLH's 366 and 367 had output growth at a higher rate than manufacturing industry as a whole. The output and productivity performance of the Electronics Computers sector (MLH 366), at 131.96% and 128.7% respectively was much higher than any other sector. There is, then, no reason, other than the performance of the Electronic Computers industry, to believe that output growth or productivity gains in the high tech sector are exceptional; on the contrary, they are lower than in manufacturing industry generally.

The Geography of High Tech in the U.K.
Poor national performance in high technology industries in the U.K. does not mean, of course, that some localities have not performed well. Indeed, in the last decade or so some areas have been popularly associated with substantial growth in high technology industries. The so-called 'M4 Corridor', from West London to Bristol and on into South Wales, is generally regarded as the most important concentration of high tech activity, with other pockets in Central Scotland and Cambridge. Despite this interest in such concentrations, few efforts have been made to measure them. Undoubtedly, data problems are largely responsible for this.

Gibbs (1983) and Ellin and Gillespie (1983) have produced maps of locational concentrations across Britain using 1971 and 1978 A.C.E. (Annual Census of Employment) statistics for functional urban regions. This work uses a grouping of M.L.H.s to define high technology industry similar to the one adopted here, and plots levels and changes in aggregate employment in these industry groups. It concludes that whilst the South-East generally shows the highest concentrations of high technology industry, there is no identifiable M4 Corridor (Gibbs, 1983, p. 21). One of the problems with this aggregate approach, also characteristic of similar American work (Glasmeier, 1985), is that it disguises the locational differences in the sectoral components of high technology industry. Also, the use of 1978 and even 1981 data to provide a 'current' picture of the location of high technology industry may be misleading. As the Berkshire County Council (1985) employers' survey, referred to later, shows, much has happened recently in the county; one suspects the same is true for central Scotland and Cambridge.

An Engineering Industry Training Board

Table 10.8: Aggregate High Technology Employment 1981, Change 1975-81, Location Quotients 1981, by selected Counties, ranked by 1975-81 Change

	County	Absolute employment 1981	Change in employment 1975-71	Percentage change 1975-81	Location quotient 1981
1	Berkshire	19,732	7,600	62.2	2.04
2	Hertfordshire	45,060	5,914	15.1	3.60
3	Clwyd	5,094	3,283	181.3	1.49
4	Hampshire	33,807	2,179	6.8	2.00
5	Surrey	18,191	1,811	11.1	1.79
6	Kent	14,650	1,679	12.9	0.96
7	West Sussex	14,694	1,669	12.8	2.03
8	Bedfordshire	7,264	1,653	29.5	1.21
9	Lothians	9,042	1,461	19.3	0.93
10	Lancashire	30,033	1,392	4.8	2.01
58	Strathclyde	23,575	-3,847	-14.0	0.88
59	Tyne and Wear	8,490	-4,032	-32.2	0.62
60	Essex	24,649	-4,342	-14.9	1.89
61	Nottinghamshire	15,860	-5,123	-24.4	1.32
62	Merseyside	14,417	-6,435	-30.8	0.86
63	West Midlands	36,500	-8,257	-18.4	1.02
64	Greater London	91,448	-17,012	-15.7	0.85
	Great Britain	640,874	-42,429	-6.2	1.00

Source: Annual Census of Employment.

(E.I.T.B) report of 1984 has produced interesting results from a survey of its members in the electronics industry. Employment is concentrated in the metropolitan counties of London, West Midlands, Manchester and Strathclyde; plus the southeastern counties of Berkshire, Hampshire, Essex and Hertfordshire. Employment change in electronics over the period 1978-1983 presents a different picture. Berkshire is the most prominent county in absolute terms gaining over 4,000 jobs with Gwent and Lothian gaining between 1,000 and 4,000. But as the E.I.T.B. (1984) paper says, it is difficult to identify an M4 Corridor, at least in electronics. Interestingly, none of the metropolitan counties grew by over 200 jobs.

Howells' (1984) study of the location of Research and Development employment (not one of our high technology producer sectors but a likely 'high technology' representative in the service sector) identifies significant concentrations in the southeast. In particular, it shows R and D employment in Berkshire and Hampshire to have risen dramatically over the 1971-1976 period. On the face of it, this would seem to give some support to the M4 Corridor phenomenon, if only for the eastern part of it. However, the A.C.E. data used by Howells is extremely suspect. Detailed analysis of the A.C.E. figures for R and D (M.L.H. 876) over the 1971-1981 period, show wild fluctuations, with a peak in 1976. Clearly, there is something wrong with these figures (possibly government research establishments were included in the data for some years and not in others?) and they should be treated with caution.

In absolute terms employment in our seven high technology industries in 1981 was concentrated in the large metropolitan areas of London (with over 91,000 jobs) the West Midlands (36,500 jobs) and Greater Manchester (27,300 jobs) and some surrounding counties, especially Hertfordshire (45,000 jobs) and Hampshire (33,800 jobs) around London, and Lancashire (30,000 jobs) north of Greater Manchester (Table 10.8 shows absolute levels for certain counties). The dominance of the metropolitan areas in absolute high technology employment reflects, of course, their size as employment centres rather than any particular concentration of high tech activity.

Table 10.8 also gives aggregate high tech location quotients for selected counties. Although again based on absolute employment in 1981, these quotients give a useful indication of the relative significance of high tech employment in each county.

311

Table 10.9 lists those counties with location quotients of more than 1.0 and high tech employment of greater than 5,000 at 1981. Hertfordshire has by far the highest concentration of high tech employment on this basis, followed by Avon and, surprisingly, Fife and Somerset.

Table 10.9: Location Quotients (above 1.0) for Counties with 5,000 or more High Technology Jobs, 1981, ranked by location quotient; base = GB.

Hertfordshire	3.60
Avon	2.32
Fife	2.24
Somerset	2.11
Berkshire	2.08
West Sussex	2.03
Lancashire	2.01
Hampshire	2.00
Essex	1.83
Derbyshire	1.83
Surrey	1.79
Mid Glamorgan	1.76
Dorset	1.69
Clwyd	1.49
Nottinghamshire	1.32
Gloucestershire	1.24
Bedfordshire	1.24
Buckinghamshire	1.15
West Midlands	1.02

As explained earlier, employment change in recent years is probably a better guide to high technology activity than absolute employment levels. Table 10.8, which ranks counties by employment change over the 1975-81 period, demotes the metropolitan counties to the foot of the table from their high ranking when assessed in absolute high tech employment terms. The counties of Berkshire, Hampshire and Hertfordshire show up well in this analysis of change, as they do in Figure 10.1, which maps aggregate high tech employment gains and losses of 2,000 jobs or more over the 1975-81 period.

The problem of analysing high technology industry in aggregate is that it disguises significant differences between individual industries. For instance, aerospace (MLH 383) is concentrated in a few counties such as Avon, Lancashire, Hertfordshire and Derbyshire. Indeed, some 90% of Avon's high technology employment was in that single industry at 1981. Similarly, in Derbyshire and Lancashire the figures were around 88%

Figure 10.1: Change in Aggregate High Technology
Employment 1975-81 (counties with + or -2,000 jobs)

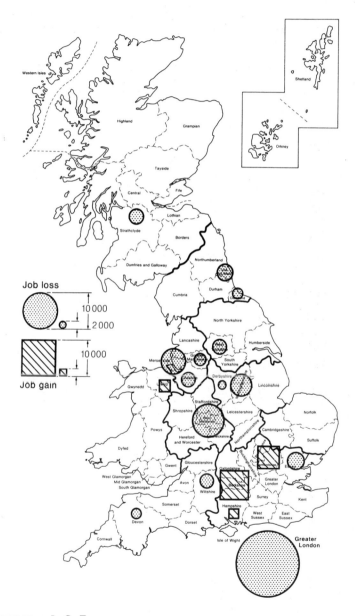

Source: A.C.E.

Table 10.10: Location quotients for counties for individual 'high technology' sectors Only location quotients above 2.0. Base = GB.

Location Quotient	MLH 272 Pharma-ceuticals	MLH 363 Telegraph and telephones	MLH 364 Radio and electronic components	MLH 365 Broadcast sound repro.	MLH 366 Electronic computers	MLH 367 Radio, radar and electronic capital goods	MLH 383 Aerospace
8.0+				Mid-Glam. Essex	Berkshire Hertfordshire		
7.5–8.0							
7.0–7.5							
6.5–7.0		Fife region					Avon
6.0–6.5							
5.5–6.0							Isle of Wight Somerset
5.0–5.5	Notts. W. Sussex						Herts. Derbyshire Lancashire
4.5–5.0	Cheshire	Cleveland Notts. W. Mids					Clwyd
4.0–4.5						W. Sussex Fife region	
3.5–4.0	Herts.	Mersey.	Fife region		Hampshire	Berkshire Lothians reg. Herts.	
3.0–3.5			Essex	Hampshire		Essex Dorset	
2.5–3.0	Kent		Beds. Mid-Glam. Wiltshire W. Sussex		Staffs.	Surrey	Glos.
2.0–2.5	Mersey.					Kent Hampshire	Dorset Surrey

314

and 77% respectively. It may therefore be more
accurate to call these predominantly aerospace
centres, and given that some 80% of the aerospace
industry's output is in defence, then defence related
high technology may be a more accurate description of
most high technology employment in these areas.

Similar distinct spatial patterns emerge in the
other high tech industries. Pharmaceuticals
dominates high tech employment in Nottinghamshire,
Cheshire and Cumbria, although in absolute terms it
is concentrated in London and Hertfordshire also.
Conversely, telegraph and telephone apparatus (MLH
363) jobs are concentrated in the West Midlands and
London, together with Merseyside and Nottinghamshire. In broadcast receiving equipment (MLH 365)
most jobs were in London, Essex, Hampshire and Mid
Glamorgan. The electronics industries remained
concentrated in the South East. Employment in
electronic computers (MLH 366) focused upon London
(16% of the national total) and Hertfordshire,
Berkshire and Hampshire (with 10 to 14% of total jobs
in the industry).

Surprisingly, some of the popularly assumed
high technology areas have very little employment in
computers, which is the sector generally assumed to
be most representative of high technology activity.
Avon, Cambridgeshire and Wiltshire, for example,
have low levels of employment (by contrast,
Staffordshire, of which we hear nothing in all the
high technology 'hype', had four times Avon's
employment in computers in 1981 and ten times
Wiltshire's).

London and the circle of home counties (Essex,
Hertfordshire, Hampshire, Berkshire and Kent)
contained most of the radio, radar and electronic
capital goods (MLH 367) jobs, although the Lothian
Region around Edinburgh and two further southeastern
counties (West Sussex and Surrey) had large numbers
of employees also. These southern counties contained
over 40% of the national jobs in the industry (60% if
London is included). Radio and electronic components
(MLH 364) employment was spread among the main
metropolitan counties and the home counties of
Hampshire and Essex.

Table 10.10 ranks counties by location
quotients (above 2.0) for each of our seven high tech
sectors, and quickly gives an impression of those
counties with high relative concentrations in each
MLH. As noted previously, employment change,
particularly over recent years, may be a better
measure of high technology activity than absolute

H.T.U.K.

levels. We have seen that nationally most high
technology sectors have been losing jobs, with the
exceptions of computers and electronic capital
goods. Hence, few counties experienced significant
employment gains over the 1975-81 period and many
lost jobs. Avon lost 5% of its aerospace jobs and
Nottinghamshire lost 30% of its pharmaceutical jobs
over the period. The telegraph and telephone industry
experienced a massive 33% national decline, but this
was not spread evenly, with the West Midlands doing
relatively well, losing only 14% of such jobs,
compared to London's fall of 29% and Tyne and Wear's
of nearly 4000 jobs. While radio and electronic
components employment fell nationally by 15% over
1975-81, Berkshire and the West Midlands each grew by
well over 1000 jobs. In the growing electronic
capital goods industry most new jobs arose in
Hertfordshire and Hampshire (over 3,500 jobs each)
and also Surrey and Kent (with over 2,000 jobs each).
In electronic computers, employment grew nationally
by 12,000 (28%), of which virtually a half went to
Berkshire. So most new jobs were again created in
counties immediately to the west of London. Indeed,
Berkshire, Hampshire and Hertfordshire alone
accounted for 40% of the employment gain in the only
two high tech sectors growing nationally (Electronic
computers and Radio and electronic components) over
the 1975-81 period. Berkshire was the only county to
experience employment gain in all seven of the high
tech sectors over this period.
    Of the other supposed concentrations of high
technology industry, Cambridgeshire fails to show up
in this analysis, with declines in employment over
both the 1971-81 and 1975-81 periods. Likewise,
Wiltshire shows a substantial decline. The
Grampians, Lothian and Tayside regions of Scotland,
on the other hand, show gains in employment, albeit
from a low base. As suggested earlier, development in
all of these areas since 1981 may be significant.

High Technology in the 'Western Crescent'
The conclusions of Gibbs (1983), Ellin and Gillespie
(1983) and E.I.T.B. (1984) seem to be borne out by
this county level analysis. The core of 'high
technology' activity, and particularly electronics,
in Britain is at the London end of the 'M4 Corridor',
in Berkshire, Hampshire and Hertfordshire. Possibly
it may be more appropriate to refer to the 'Western
Crescent' or the 'West London Crescent'. Areas
further west, particularly Wiltshire and Avon, which

316

are popularly believed to feature strongly in the M4
phenomenon, show little particular advantage.
Likewise, South Wales generally shows losses rather
than gains in our high technology sectors.

Analysis of A.C.E. data at finer geographical
areas than the county level is necessarily limited
because of confidentiality restrictions. Townshend
(1986) has carried out an assessment of high tech
employment at this level across the U.K. using Travel
to Work Areas. He finds a number of expected
concentrations of growth; in the Reading area, for
example. But he also finds significant localised,
high tech growth in some less obvious areas, such as
Preston, Macclesfield and Dundee. As part of the
Reading project, an analysis was carried out using
Functional Urban Regions (F.U.R.s), which are
similar to T.T.W.A.s. This showed that, in the south
of England, the general pattern of change revealed by
the county level analysis is confirmed. However, the
centres of gains and losses in employment can be
identified more precisely (see Figure 10.2). Hemel
Hempstead, Bracknell, Reading and Portsmouth are the
major centres of growth. The major centres of decline
are London, and, perhaps surprisingly, Harlow,
Cambridge, Gosport and Swindon. In South Wales,
Swansea has a small but significant gain in an area
that otherwise shows general high tech decline.

One major problem with all of the sub-national
analysis so far is that it stops at 1981. This is
because no official employment data are available
beyond that date. This is unfortunate because clearly
a lot of the high tech activity we are interested in
has emerged since 1981. Indeed, a visit to Swindon
for example, makes the picture painted by employment
change to 1981, which shows substantial high tech
decline, difficult to believe. A lot must have
happened since 1981.

The only available post-1981 employment
information for counties or finer areas comes from
surveys. Fortunately, some local authorities in the
'Western Corridor' have carried out their own
employment surveys in the last year or so.

Hampshire County Council (1984) has undertaken
a survey of specifically high technology firms. Using
a broader definition of high tech than that referred
to above (it includes certain service sectors), the
survey identifies 46,400 jobs in the County (10% of
the total) with 29,500 in South Hampshire and 12,500
in North East Hampshire. It is in this latter area,
immediately south of Berkshire, that the greatest
rate of increase has taken place. Just over a half of

Figure 10.2: Change in Aggregate High Technology
Employment for Functional Urban Regions, 1975–81
(Source A.C.E.)

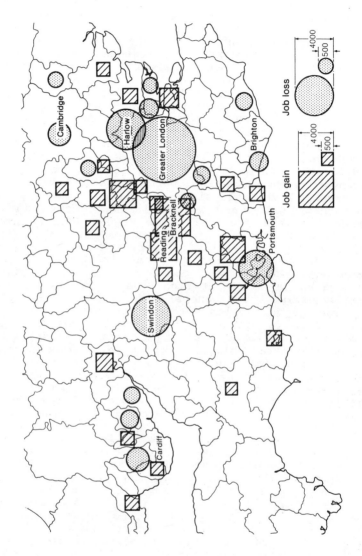

the total high tech employment was found to be in the electronics sectors (see Table 10.11). The occupational structure of high tech industries, as shown in Table 10.12, shows the very high proportion of total jobs taken by those in managerial/profess-ional and technical occupations. Males predominate in this category, taking 92% of the jobs.

Table 10.11: Types of high technology industry in Hampshire, 1984

Industry	No of firms	Jobs
Instrument engineering	39	5,884
Other engineering	17	3,028
Electronics	93	22,779
Aerospace	10	3,231
Business/professional services	23	1,412
Research and development	5	6,458
Other	4	1,229
Total	191	44,021

Source: Hampshire County Council (1984)

Table 10.12: Occupations by sex in high technology in Hampshire, 1984

Occupation	Males	females	Total	per cent females
Manag/prof/tech.	13,659	1,216	14,875	8
Other non-manual	6,901	5,337	12,238	44
Skilled manual	6,417	437	6,854	6
Semi/unskilled manual	5,311	5,419	10,730	50
Total	32,288	12,409	44,697	28

Source: Hampshire County Council (1984)

The most comprehensive of the surveys carried out is that of Berkshire by Berkshire County Council (1985) with Newbury District Council, Bracknell District Council and with some help from the University of Reading. With the exception of some small sectors, this attempted a 100% survey of employers. The results of this survey are particularly interesting for our assessment of high tech industry because Berkshire consistently emerges

Table 10:13: High Technology Employment in Berkshire, 1984

	Firms	Manuf (a)	Whole (b)	Office (c)	Total	% of emp*	Estimate of total	Av. size of firm
Newbury	41	981	84	97	1,162	5.8	2,000	28
Reading	77	1,871	1,383	695	3,949	12.3	6,000	51
Wokingham	51	1,057	464	286	1,807	14.6	4,000	35
Bracknell	34	3,408	329	3,187	6,924	30.6	9,000	203
Wind/Maid.	65	193	489	1,164	1,846	8.1	3,000	28
Slough	59	964	1,500	966	3,430	12.9	6,000	58
Berkshire	327	8,474	4,249	6,395	19,118	14.0	30,000	58

(a) Manufacturing
(b) Non-manufacturing operating in industrial/warehouse premises, usually engaged in a mixture of wholesale distribution, servicing, some R&D, adminstration, software production
(c) Non-manufacturing operating in office premises, usually engaged in software production or sales, and marketing of electronic products or pure R&D

* of survey private sector employment

Source: Berkshire C.C. (1985)

in the analysis presented earlier as a centre of this
activity. There is reason to believe that the county
has the fastest rate of growth of high technology
industry in Britain at present and, hence, is at the
'leading edge' both of these developments and of
their social and economic consequences.
     Table 10.13 lists high technology employment in
Berkshire Districts by a three-way classification of
job type. The manufacturing sector includes firms in
the electronics fields corresponding roughly to the
previously discussed definition. In addition some
biotechnology firms were included. The wholesale
firms were those in warehouse type premises engaged
in wholesale distribution, plus a mixture of
servicing, R&D, administration and software
production. The office firms were in office premises
usually engaged in software production, sales,
marketing or pure R&D for electronics products. Hence
the wholesaling and office groups would not fall
under our MLH industries discussed previously,
although the manufacturing would. The overall
estimate of approximately 30,000 high technology
jobs in the county compares with a figure of
approximately 22,000 estimated from a County Council
survey conducted at the end of 1981. Thus, over a two
year period high technology employment as defined by
the County has increased by approximately 8,000. This
compares with an A.C.E. growth of 7,800 over the ten
year period from 1971-81 and shows, as suggested
earlier, that the post-1981 changes may be very
significant.
     Bracknell District has the lion's share of the
high tech jobs in Berkshire, with the much vaunted
Newbury having the lowest level amongst the Districts
(Table 10.13). Interestingly, the well-established
industrial areas of Bracknell and Slough have firms
which are larger, on average, than those in other
Districts. These larger firms, particularly in
Bracknell, seem to be functioning as offices, R&D and
prototype development units rather than as
manufacturing plants. Manufacturing features most
strongly, as a proportion of total high technology
employment, in the more recently industrialised
areas of Newbury and Wokingham to the west of the
county. In these Districts, manufacturing is in small
establishments, often producing customised electronic components. Bracknell has by far the largest
proportion (30.6%) of its surveyed employment in high
technology industries, with Newbury (5.8%) a long way
below the county average (14%). If we include public
sector employment, then high technology accounts for

approximately 10% of county employment; this compared with a national figure of 3.84% (at 1982).

Table 10.14: High technology industry in Berkshire; Function of Site by Activity (Percentage of Firms)

	manuf	whole	office	overall	all firms
No other site	63%	29%	32%	40%	50%
HQ site	13%	19%	22%	18%	11%
Branch, HQ in UK	15%	20%	20%	19%	34%
Branch, HQ abroad	9%	32%	26%	23%	4%

Note: does not include Bracknell and Newbury

Source: Berkshire County Council (1985)

Table 10.15: High technology industry in Berkshire; Function of Site by Age of Firm (Percentage of Firms)

	employees				
	1-9	6-10	25-99	100+	overall
No other site	53%	44%	31%	11%	40%
HQ site	7%	18%	26%	36%	18%
Branch, HQ in UK	13%	17%	20%	39%	19%
Branch, HQ abroad	27%	21%	23%	14%	23%

Source: Berkshire County Council (1985)

Some 40% of the surveyed high tech companies operate on single sites, with the figure being as high as 63% for manufacturing firms. 23% of high technology firms are foreign owned, compared to only 4% for all firms in the survey (Table 10.14), foreign ownership being much higher amongst the warehousing and office activities than in manufacturing. This highlights the distribution role played by the Berkshire establishments of many foreign firms. Table 10.15 shows that a large proportion of single site firms (53%) are small, having 9 or less employees. Surprisingly 27% of the smallest (1-9 employees) firms are foreign-owned. These are likely to be again distribution centres or offices for foreign manufacturers. Table 10.16 relates the age of firm, in years since establishment, to the function of sites. The foreign-owned element is consistent over the age range, varying between 19% and 27%, suggesting no sudden influx of firms in recent years.

The relatively recent growth in high technology companies is shown by their length of time in the county. These companies generally have been established in the county more recently than other companies. 24% of all companies have been in the county for less than 5 years compared to 49% of high technology companies.

Table 10.16: High technology industry in Berkshire; function of site by age of firm

	Age of firm			
	1-5	6-10	11-24	25+ years
No other site	43%	45%	32%	*
HQ site	18%	21%	15%	*
Branch, HQ in UK	12%	14%	28%	*
Branch, HQ abroad	27%	19%	24%	*

* sample size too small

Source: Berkshire County Council (1985)

Table 10.17 shows that, as expected, high technology companies employ a greater proportion of their workforce in professional and technical capacities than companies overall, and a lower proportion in skilled and unskilled manual work. This provides further confirmation of the well-recognised problem that growth in high technology sectors will not provide direct compensation for jobs lost in traditional manufacturing. Even in areas like Berkshire, with relatively slower decline in these traditional sectors and growth in new sectors, this is still a problem.

Table 10.17: High Technology Industry in Berkshire: Employment by Skill Structure, All Employment and High Tech Employment

	All	High Tech
Managerial	12%	13%
Professional/tech.	20%	42%
Clerical	26%	25%
Skilled manual	13%	8%
Unskilled manual	29%	12%

Source: Berkshire County Council (1985)

Explaining the Development of the Western Crescent
So far, then, we have identified a concentration of high technology industry to the west of London, albeit taking the form of a 'Crescent' rather than a long, continuous 'Corridor'. We have also identified certain important characteristics of this concentration: it has witnessed a relatively high growth in high tech jobs, many of them in R and D, administration and marketing; it has a broad base of high tech industries, while being particularly strong in electronics; and it has a high level of foreign ownership amongst its high tech companies. The question remains, however, as to why this concentration has occurred in this form, in this place, at this time.
A lengthy explanation is attempted in Hall et al (1986 forthcoming). Here just the main components of this explanation will be given.

The Spatial Division of High Technology Labour. The survey results presented above immediately go some way to giving us some form of explanation; in particular of the role that Berkshire plays in the wider world of high technology industry. A popular conceptual framework for helping to explain the changing spatial nature of industrial activity, at international and national levels, in recent years concerns 'spatial divisions of labour' (Massey, 1984). According to this conception, the increasing tendency for industrial development to be spatially uneven can be explained by modern capital continually looking to arrange the spatial division of its activities in such a way that reaps the greatest benefits. The consequences of this can be felt at international, national or at local levels. This conception, allied to the survey results, may help to explain Berkshire's role in the 'spatial division of high technology labour'.
At the global level, the main actor in this view is the multinational corporation, viewing virtually the whole world as its own space economy over which it can manipulate locations to its greatest advantage. Typically, the outcome of this might be a hierarchial division of activities, with overall control maintained in the parent country (and city), possibly divisional headquarters in appropriate developed countries, and production elsewhere, typically in a low-wage developing country (Hymer, 1975). Thus, the control functions of the corporation - administration, R and D., marketing - will usually

remain in the parent nation.

Hymer's (1975) particular conception of the spatial division of labour can be characterised diagramatically as in Figure 10.3.

Figure 10.3: Hymer's stereotype of the spatial division of labour

Level of corporate hierarchy	Type of area		
	major metropolis (e.g. New York)	regional capital (e.g. Brussels)	periphery (e.g. South Korea or Ireland)
1. Long term strategic planning	A		
2. Management of divisions	D	B	
3. Production, routine work	F	E	C

Source: Hymer (1975)

Within this framework, the expected pattern of activity for multinational corporations in an international one linking A-B-C. For multi-plant national companies, operating only within a particular country, we might expect a B-C relationship, with headquarter functions in one region, typically the South East in the U.K., and branch plants, carrying out more routine production in a relatively low-wage, high government incentive region.

Much of the debate concerning spatial divisions of labour concerns the consequences of corporate activity being organised in this way. The most obvious concern is the loss of local - be it a developing nation state or a depressed region in a developed country - control over local industrial activity. Thus, a local economy may be at the mercy of decisions made thousands of miles away. There is good reason for such concern, for there is ample evidence that the first to suffer when recession hits multinational companies will be peripheral branch plants.

Some researchers have considered the specific role of high technology companies in this new international division of labour (Castells, 1985, Sayer and Morgan, this volume, Henderson and Scott, this volume). Interestingly, high technology has a major role in two ways. Firstly, because the manufacture of high technology products is a major, and relatively new industrial activity, large corporations are still establishing their preferred global spatial strategies. Thus, for example, in recent years the electronics sector has been subjected to this treatment. Secondly, the very products of high technology industries actually facilitate the development of corporate spatial strategies. With modern computing and telecommunication devices, corporate activities can be spread across the globe, but still remain under direct and immediate control of headquarters.

The interesting question from our point of view, is the role that the Western Crescent and particularly Berkshire plays in this international high technology space economy. Is it performing a B role in Hymers's conception, acting as a divisional headquarters for overseas companies, or is it a peripheral, production location of type C? How does it fit in with other high technology centres in the U.K., such as South Wales, Central Scotland or Cambridge?

In the Crescent, high tech manufacturing does appear to be less significant than in other supposed concentrations of high technology activity. There is manufacturing, but much of it consists of short-run, high value, customised products or, indeed, prototype production. There is little of the long-run volume production that we now associate with, say, South Wales and Central Scotland. There is a considerable amount of research and development and administrative activity. Also many companies, particularly those from overseas, have marketing and distribution functions in the area. To a certain extent the area conforms to standard notions of the spatial division of labour, with Berkshire serving a 'control' function over areas with a production role. Indeed, this role appears to have become even more marked in recent years. The Berkshire County Council (1985) survey distinguishes between the functions of electronics companies which have been in the county for five or more years and those established during the last five years. As Table 10.18 shows, those companies established in recent years have a predominantly office function - which we associate with administration and R and D - while older electronics companies are still performing a significant manufacturing activity in the county.

326

Table 10.18: Functions of Berkshire Electronics Firms by Date of Establishment

	Established more than 5 years	Established during last 5 years
Manufacturing	48%	21%
Office	27%	49%
Wholesale	25%	30%

Source: Berkshire County Council (1985)

There are, however, many exceptions to the standard notions about the spatial division of labour. Some firms, particularly small ones, operate only in Berkshire. Other larger firms have a corporate structure which defies the assumed control/production split. Racal, for example, which is a major UK electronics company heavily represented in Berkshire, consists of a set of semi-autonomous companies ('Racal - Suffix'), none of which is very large, and in which just about all functions are located at the one establishment.

The particular current form of the high tech activity in Berkshire, and in the other Western Crescent counties, may have important implications for the future. One of the features of the control/production split hypothesis is that areas of production may suffer most in periods of retraction. This is likely to be the case as much in high tech sectors as in any other. It may, in fact, happen without retraction if, as is suggested, production techniques in sectors such as computer manufacture are amongst the least automated in industry. Increased automation in the electronics sector generally is much more likely to have an adverse effect in Central Scotland and South Wales than in Berkshire. Advantaged areas such as Berkshire may, then, have the additional benefit of being relatively durable when times get harder.

Historical Origins. When the Reading project started, a number of working hypotheses were posed. These are outlined in Breheny, Cheshire and Langridge (1983). In the event, certain features of these hypotheses were confirmed, but in a number of respects the hypotheses were found to be inadequate. In general, the explanation proves to be more complex than expected and has deeper historical origins than initially assumed.

The origins of many of the high technology industries and firms now in the Western Crescent can be traced to the electrical industry in west London, although why large parts of that industry originated in London is uncertain. For whatever reason, a concentration of production of the high technology products of the day - radios, gramophones and televisions - existed in west London in the 1920s and 30s.

Even as late as 1951, according to the Census of that year, 109,437 people in the Greater London conurbation, or 46% of the national (England and Wales) industry total, were employed in the electrical wire, telegraph and telephone apparatus, wireless apparatus and wireless valves industries. According to Keeble (1976) both professional electronics (computers, aerospace and military) and consumer electronics (radios, cathode ray tubes etc) were heavily concentrated in Greater London during the 1950s due to environmental advantages for innovative activities, market accessibility (for both government and other consumers) and skilled labour availability.

From the 1950s these industries underwent relative dispersal, although in absolute terms employment still grew in London until the early 1970s. However, while there was considerable growth in production jobs in these industries in the development areas, much of the research and development remained in the South East due to the need to attract large numbers of graduates. From 1959-71 the major share of growth in electronics remained in the South East, particularly Berkshire, Hampshire and London (Keeble, 1976). Much of this growth around London seems to have been in firms from the 'seedbed' around north west London, such as Standard Telephone, G.E.C., Solarton, Racal, and M.O. Valve (Martin, 1966, 101, 101-102). Solarton Electronics moved out to Farnborough in Hampshire, and Racal moved from Isleworth to Bracknell. Similarly a large number of other firms moved out from near the Great West Road to Bracknell in the 1950s, mostly due to high costs in the former locations and the attractions of the new towns (Brown 1966).

The pattern of dispersal of these and other electronics companies seems to have been influenced from the 1930s onwards by increasing contact with Government Research Establishments (G.R.E.s), and particularly Defence Research Establishments

(D.R.E.s). The spatial distribution of G.R.E.s in southern England is discussed in some detail in Hall et al (1986 forthcoming), but for our present purpose it is sufficient to say that these have, since the 1930s, had a distinct bias towards the area west and south-west of London. There is little doubt that the decentralisation pattern of a number of electronics companies, already nearby in west London, was influenced by the need for ready access to D.R.E.s , such as the Royal Aircraft Establishment at Farnborough and the Atomic Weapons Research Establishment at Aldermaston. Many companies which had previously worked solely on products for civil markets, found themselves working in the build up to and during World War II on the production of military equipment, such as radios and radar. Many electronics and aerospace companies maintained their links with the military after the war and continued to receive contracts, particularly when the Cold War period of the 1950s intensified the demand for sophisticated electronic weapons and equipment. Thus, the close contractual and spatial relationship between electronics and aerospace companies and D.R.E.s that we see today was established some considerable time ago.

Infrastructure Investment and Planning Policy.
Pursuing still further the historical origins of the growth of high technology industry in the Western Crescent, we can consider usefully the role of infrastructure investments and planning policy.

Communications have always been vital for the development of the area from London to Bristol. Swindon was a small market town before the repair shops were built for the Great Western Railway in 1841. Slough was the site of a major transport depot of the War Office during the First World War, sitting as it did on the Great Western Railway and the Bath Road. After the war the site was developed as one of the first private industrial trading estates and the town grew from 27,000 people in 1921 to 63,000 in 1939. Meanwhile, Windsor just a few miles south, but not on a major road or railway, stagnated in the 1920s and lost population in the 1930s (Hall et al, 1973). Further out, residential villages and towns such as Ascot and Wokingham flanked the railways and these, as well as the towns on the mainlines such as Maidenhead and Woking, shared in the explosion of growth in the whole London region between the wars. To the south, Aldershot was developed as a military

base during the Crimea War, and later the Royal
Aircraft Establishment developed at Farnborough.

Of crucial importance to the subsequent
development of the Western Crescent was the decision
at the end of World War II to establish Heathrow as
the country's major international airport. As will be
demonstrated later, ready access to Heathrow has
proved to be a major attraction of the area for high
tech companies. During the 1970s, the opening of the
M4 motorway westwards from London, past Heathrow,
increased the attractiveness of locations west of the
airport. As will be demonstrated later, the M4
motorway, and the access it provides to the national
motorway network, is regarded by high tech companies
as a major attraction of the Western Crescent
location.

Planning policy has played a significant role in
the development of the Western Crescent, albeit a
rather disjointed one. Abercrombie's Greater London
Plan (1944, 161) suggested further development of
Slough, restriction in Windsor, and importantly, the
development of a new town at White Waltham west of
Maidenhead. The new town was eventually built at
Bracknell, about five miles south. The planning
system also had considerable impacts on the expanding
electrical engineering and electronics industries.
The post-world war II 'ceilings' on factory expansion
helped push firms or plants short distances (15-50)
miles out of northwest London. In addition, due to
their good export record, and possibly because of
their importance for defence, electronics companies
found it easier than other companies to obtain
Industrial Development Certificates in the
prosperous South East and in new towns close to
London (Keeble, 1968, 18). To generalise, it could be
argued that IDCs pushed what were to become the
declining sectors into development areas, while
allowing the advanced and future growth sectors to
remain in the South East.

The designation of the Reading-Wokingham area
as a major growth zone in the Strategic Plan for the
South East of 1971 (Joint Team, 1971) has no doubt
been important in facilitating some of the recent
growth in that part of the Western Crescent.
Interestingly, however, it may well be that long-
standing policies of restraint over much of the
Crescent area by local authorities may be just as
important in accounting for high tech growth as any
localised growth policies. These restraint policies
have had the effect of retaining much of the area's
very pleasant environment. A peculiar feature of high

tech companies is that they often seek rural or semi-urban locations with attractive physical environ-ments. Thus, the Western Crescent, which also has strong cultural attractions (private schools, horse-racing, access to London) for the affluent, has an immediate publicly preserved attraction for high tech companies. An interesting question to consider at the present time is whether recent growth, of both housing and industry, will spoil the very environmental attractions that have given rise to much of this growth?

The Western Crescent, then, has had the advantage, over many years, of substantial public-sector support - major infrastructure investments, including G.R.E.s, roads, and Heathrow, and positive and negative planning - which has stood it in good stead in attracting modern industrial development. We must conclude, however, that the overall result of this public sector support is largely accidental. Major investment decisions have been made individually - by the Ministry of Defence, the Department of Transport and the Department of Trade - not as part of any overall strategy aimed at creating a concentration of modern industrial development.

Nevertheless, attempts to explain the spatial pattern of high tech development solely in terms of the motives of the private sector - which is the dominant line in this field - are likely to be misguided. Certainly the public sector has played a major role in the Western Crescent in giving the area, over a long period, a combination of characteristics that have proved attractive to high tech industrialists. The point is not only important for explanatory purposes, but also for prescriptive ones. In studying carefully the past role of the public sector in helping to bring about, however unwittingly, conditions for local growth, we might gain some idea of how public intervention might aid more depressed localities.

Survey Evidence. In order to help our explanation of the development of the Western Crescent, an interview survey covering electronics based firms in Berkshire was carried out during 1984. Although a wide range of questions was asked, the main ones concerned the reasons why the firms had chosen their present location and what were currently perceived to be the advantages and disadvantages of the area. Of the forty-four establishments (each representing a different firm), 13 had manufacturing as their main function, 11 had R&D functions as well as manufacturing, 13 were concerned primarily with distribution, sales and servicing and offices (two of which did some manufacturing also), 3 were software or training 'plants', two were R&D centres and two were offices of R&D centres. Each establishment carried out a number of functions.

When considering why firms have located in Berkshire, it is useful to distinguish new single plant establishments from branches of foreign or U.K. firms. Twenty-four of the firms interviewed fell into the first category. The reason for the initial location of the new firm in 21 cases was simply that it was within commuting distance of where the founder lived when he (unfortunately, all 'he's') finished his previous employment. Hence, if one wants to understand why a new firm starts in a general location (as opposed to a specific site) then residence is important. Our sample obviously excluded those leaving the area to start up, but should have included those moving to the area.

Typically, the founder of a new company in the 'Crescent' previously worked in London, particularly west London, for a large company (say, Plessey, Thorn-EMI, or G.E.C.) and commuted to work from a residence further west in, say, Berkshire. Upon leaving his large employer, which we may term the 'spawning' company, the employee naturally looked to set up his new 'spawned' company (Breheny and McQuaid, 1984) near to home, in, say, Reading, Bracknell or Wokingham. Thus, this spawning process has had the distinct effect of locating small new firm formations west of the larger established locations.

Interestingly, this spawning process has specific characteristics. Firstly, the founders of companies in the survey came solely from private sector companies. In the working hypotheses posed by Breheny, Cheshire and Langridge (1983) it was assumed that a good source of spawned companies would be the G.R.E.s, with government researchers and scientists seeking to use their talents in the private sector. In fact, not only did no founders of companies come from G.R.E.s, but remarkably few employees in such companies had come from these Establishments. The second surprising feature of the spawning process was that the larger spawning companies that produced the founders of new spawned companies were invariably British. The large number of overseas companies, particularly American ones, in west London and the Western Crescent generally, did not produce a single

founder of any of the new high tech companies in our
survey.

Of the founders of new firms in the survey nine
had worked immediately beforehand for a large (over
200 employees) high tech firm west of London, five
for a high tech firm of under 200 people and five for
a non-high tech firm. Of the remainder one gave no
response and four were freelance or contract workers
(two for large high tech firms). The location of
these large firms west of London helps to explain the
growth of new firms further west. Two of the three
firms that did not locate close to the founders' home
picked Reading because of excellent communications
(especially rail) with London, and the other chose
the area because of labour supply and cost (this firm
was founded in 1959). Although many of these single
site firms had moved or were considering moving,
invariably the move or intended move was local, so as
not to lose skilled workers.

The reasons were for the initial choice of
location by foreign or U.K. branch plants were
difficult to discover, particularly as they were
usually made many years before and it was not clear
who had made the final decision. The foreign branches
had generally been located in the South East due to
closeness to London and to Heathrow and to a wide
choice of transatlantic and European flights. The
foreign owned firms are generally late arrivals to
the Western Crescent. Typically they will have
established their first base in the 1970s. As
explained earlier, the structure of foreign
electronics firms in the U.K. is often organised with
the R&D and office functions in the South East and
production elsewhere. For instance, the four sites of
Hewlett-Packard in Berkshire (at 1984/85) cover
administration, R&D, sales, and servicing.
Production is carried out at South Queensferry near
Edinburgh, and near Bristol. Similarly, Honeywell
have a development establishment in Hertfordshire,
their two section-head offices in Brentford and
Bracknell, but production together with a design and
development group in Scotland. While Digital had 800
employees at their UK head offices in Reading at the
time of the interview (housing sales, service,
administration, field service logistics, and
computer services design and manufacture), their UK
production plant employed more than 500 people at Ayr
in Scotland. Six other Digital sites in Reading for
training and associated activities employed a
further 340 workers. Another 250 people worked on
customer support and sales at Basingstoke,

Table 10.19: Perceived Advantages/Disadvantages of a Thames Valley Location

	Percentage of Firms					
	Type 1 Single-site		Type 2 + 3 Multi-site		Total Firms	
	−	+	−	+	−	+
Labour						
Local availability of:						
Adminstrators and Managers	5	−	−	17	−	8
Professional & Scientific Staff	5	14	6	28	5	20
Clerical Staff	−	9	6	17	3	13
Skilled & Supervisory Staff	14	18	11	17	13	18
Semi- & unskilled Staff	14	5	6	17	10	10
Easy to attract labour	5	5	6	17	5	10
Cost of labour	9	−	6	6	8	3
Environment						
Housing Cost	9	−	44	11	25	5
Housing availability	9	9	−	17	5	8
Cultural/Recreation Facilities	−	9	−	6	−	8
Pleasant Place to live	−	23	−	17	−	20
Good environment (not specified)	−	5	−	22	−	13
Social relations with others in the same industry	−	−	−	17	−	8
Communications						
Heathrow Airport	−	77	−	72	−	75
Other Airports	−	9	−	11	−	10
M4 Motorway	−	73	−	50	−	63
Other Motorways and Major Roads	−	36	−	44	−	40
Rail Network	−	23	−	22	−	23

Table 10.19: continued

Agglomeration Characteristics	Percentage of Firms					
	Type 1 Single-site		Type 2 + 3 Multi-site		Total Firms	
	-	+	-	+	-	+
Access to:						
Govt. Research Establishments	-	5	-	17	-	10
Universities/Higher Education	-	-	-	11	-	5
Local Business Services	-	5	-	11	-	11
Local Suppliers	-	23	-	6	-	18
Local Customers	-	36	-	39	-	40
Exchange of Ideas with Others in the Industry	-	-	-	6	-	3
Near National Government Depts.	-	-	-	11	-	5
Close to other parts of Firm	-	-	-	22	-	11
Access to private R&D Facilities	-	-	-	6	-	3
Premises						
Suitable Quality	5	23	-	44	3	33
Suitable Availability	5	41	-	39	3	40
Suitable Rent Levels	18	9	11	17	15	13
Suitable Rate Levels	9	-	6	6	8	3

Note: '-' signifies disadvantage of area; '+' signifies advantage.

Source: Interviews

335

(unpublished Digital memo, January 1984). Most of the large U.S. firms seem to follow similar types of spatial distribution of functions within the U.K.

The branches of U.K. firms each had differing reasons for locating in the Crescent. In the case of three large firms, the closeness of government defence research establishments was important (see below). In addition the 'push' from London due to expansion or reorganisation and the 'pull' of the new town of Bracknell played a role in the location decisions of certain companies. It is not possible to distill a specific set of reasons for the large firms locating in the Crescent although the role of outmigration from London, the attraction of Heathrow, the perception of the area as a 'pleasant place to live' and the availability of skilled labour were recurrent themes.

The firms were asked what they considered to be the advantages of the eastern part of the 'M4 Corridor' (or the 'Thames Valley') compared to other parts of the country. This approach avoided the problem of ex-post rationalisation by the interviewee when presented with a list of location factors. The main factors of importance for firms were labour, environment, communications, agglomeration economies and premises.

A pool of professional and skilled labour was often seen as a major attraction of the area. Twenty per cent of firms specifically stated that the availability of professional and scientific staff was an important advantage of being located in this area (Table 10.19). While there was a shortage for many firms of types of engineers and other professional workers, this was seen as a national, not a local problem. Branch plants (i.e. establishments that are part of a multi-establishment firm) put a greater emphasis on the availability of professional workers (28%) compared to single plant establishments (14%), although this may reflect the demand for larger numbers of such staff by the generally larger, R and D orientated branch plants. The advantages of the area for obtaining skilled production and supervisory staff was equally important (about 18%) for each type of firm. However, some 14% of single site firms and 11% of branch plants (mostly manufacturing plants) considered it difficult to recruit skilled production workers in this area.

The environment has been considered important in the development of Cambridge and Silicon Valley (Segal Quince, 1985; Saxenian, 1983). In the

Berkshire study some 30% of firms considered the environment to be an advantage of the area, although only two firms mentioned it as having any significance in their location choice. The good environment was generally expressed as 'a pleasant place to live' although 5% of firms said culture/recreation facilities were also important. On the other hand 25% of firms considered high housing costs as a disadvantage of the area, often leading to difficulty in bringing in technical and skilled workers from other parts of the country.

Three-quarters of firms cited Heathrow as an important advantage of the area. Surprisingly, slightly more single site firms saw it as advantageous (77%) compared to multiplant firms (72%). Other airports (only 10%), were not seen as very important. The M4 motorway was important for 63% of firms (73% of single site firms and 50% of multi-site ones), while some 40% of firms thought other major roads to be an important advantage, particularly for north-south travel. The railway was important for nearly a quarter of the firms (23%). The main reasons for the importance of air communications were the movement of output and inputs to and from overseas (over 40% of firms each) and movement of personnel (including a number frequently travelling to Scotland). The road network was especially important for moving output throughout the country (43% of firms) and particularly for site accessibility for customers (48% of firms). Generally road communications were more important, and air movement of goods less important, for single site compared to multi plant firms (despite Heathrow being more often cited by the former). While these questions referred to the general location (eastern 'M4 Corridor') road networks are likely to be much more more important for local site selection.

By far the most important agglomeration advantage of the area was access for local customers (40% of firms). This was equally important for single and multiplant firms. For small firms, access to local suppliers was quite important (23% of firms), though this was less so for multisite firms (11%) who received supplies from other branches of the company or who expected the suppliers to come to them (possibly reflecting the large purchasing power of many of these firms). For a number of the multibranch firms (22%) the location of other parts of the firm in the UK was important. Surprisingly, access to business services was considered advantageous in this area by only 8% of firms, perhaps as they were

perceived to be readily available in most parts of
the U.K. Access to local universities - important in
Cambridge and Silicon Valley - were thought important
by only 5% of firms. This helps to indicate the
different nature of high tech development in the
Western Crescent compared to those other areas.

The area was generally considered advantageous
for the availability of suitable premises by about
40% of both single and multiplant firms but only at a
price. Eighteen per cent of single site and 11% of
multi-site plants thought the area disadvantageous
in terms of rent levels. Rate levels (local property
taxes) were thought excessive in this area by only 8%
of all firms. Of course, the rate levels are
considerably below those in London, so rates may be
considered moderate by companies with experience of
operating in the capital.

Table 10.20: Factors aiding or hindering the
Expansion of Businesses

	Percentage of Firms					
	Type 1 Single-Site		Type 2+3 Multi-Site		Total Firms	
	+	−	+	−	+	−
Demand for Output	36	0	33	0	35	0
Labour Shortage	0	36	0	22	0	30
Labour Costs	5	5	6	6	5	5
Working Practices	0	0	6	6	3	3
Financing Constraints	5	18	0	22	3	20
Site Constraints	0	27	0	11	0	20
No Answer	27		50		38	
(No of Firms)	(22)		(18)		(40)	

Source: Interviews.

In addition to these questions about the merits
of a Thames Valley location, the interview survey
also covered factors helping or hindering the
expansion of firms. Virtually the only important
factor helping expansion was increases in demand for
the firms' output. This was mentioned by 36% of
single and 33% of multiplant firms (Table 10.20).
Conversely, 36% of single and 22% of multiplant firms
were hindered in growing by labour shortages, and
only 5% overall by labour costs. When firms not
answering this question are excluded, the labour
shortage figures are 50% and 44% respectively. This
must lead to questioning of policies aimed at
reducing labour costs rather than improving labour

skills for helping firms expand employment. The two other main hinderances to firms hoping to expand were site constraints (27% of single and 11% of multi-site firms) and financing constraints (18% and 22% of firms respectively). Interestingly, only one firm considered working practices a hinderance. In general then, expansion of the firms depended upon rising demand, while such expansion was hindered by financing constraints, site constraints (for single site firms) and labour shortages, but not usually by labour costs or practices.

The Role of Defence Spending. Surprisingly in retrospect, the original working hypotheses (Breheny, Cheshire and Langridge, 1983) of the M4 study did not focus on defence spending as a significant factor in the development of the Western Crescent. However, the interview survey quickly revealed this to be a major factor. In fact, the history of high technology industries has been closely tied to defence spending in many countries. The major early development work on computers and the microchip in the U.K. were carried out at government research establishments or in firms carrying out military research (see for instance: Plessey 1983; Lavington, 1980; Kaldor, 1982; and for more recent developments, Barnaby 1982). In 1983 over half of all government funding for R&D went on military purposes (Cabinet Office, 1984). Many of the most advanced technological developments and most advanced high technology firms are closely tied to defence spending. Some 20% of the output of the UK electronics industry and 50% of aerospace output go to the Ministry of Defence. The MOD itself argues that it supplies much of the R&D support for industries such as electronics, aviation control systems and marine technology, without which these industrial sectors 'could not function effectively' (Ministry of Defence, undated). Clearly then, it is difficult to analyse high technology industries generally without taking this into account.

At the level of individual firms, dependency on defence is often considerable. In 1984, out of a turnover at British Aerospace of £2.5bn, some £2.0bn was from defence sales, with £900m accounted for by the MOD. Of the sales to British customers by Plessey in 1983, some 30% were to government departments, especially the MOD and many of their exports were of military hardware (Plessey, 1983). Arms production as a percentage of total sales are high in a number

Table 10.21: Defence and Regional Expenditure, by Regions, 1977/78 (£ million)

	Defence Total	Regional (except Defence)	Defence Procurement Contracts	Regional Aid Grants
South East	2518.2	14022.2	1175.6	–
South West	1107.2	3010.1	410.2	7.8
East Anglia	243.0	1274.0	111.7	–
East Midlands	382.7	2634.0	213.2	1.4
West Midlands	346.8	3720.8	189.2	–
North West	382.1	5372.3	242.8	57.6
Yorkshire/Humberside	244.9	3759.2	61.9	26.7
Northern	242.1	2824.6	174.2	142.3
Scotland	497.1	5246.9	195.0	113.1
Wales	138.3	2481.2	25.6	74.4
Northern Ireland	160.6	1797.9	27.9	50.1

i   Defence procurement contracts include: purchase of defence hardware for air, land and sea systems; clothing and textiles, liquid fuels, lubricants, etc.; research; and contract repair of ships and vessels.

ii   Expenditure for regions includes all government expenditure (central, local and nationalised industries) for the benefit of the region, but excluding expenditure for the benefit of the country as a whole, such as defence, prisons and foreign affairs.

iii   Regional Aid Grants include Regional development Grants (paid under the Industry Act, 1972) and selective assistance grants. Note that the East and West Midlands together received £0.5 million in selective assistance grants which are not included in the table. For Northern Ireland the grant figures are for Investment and capital grants and Industrial Development Grants.

Sources: Short, 1981a and 1981b; Regional Statistics, 1978, 1979.

of other well known companies along the M4 corridor. For Racal this figure was 39% in 1984, for Ferranti 51%, for Plessey 37%, and for Westland 84% (Street and Beasley, 1985). In each case a large proportion of this defence work is for the MOD, but defence exports are also very significant. In 1984/85 defence sales from the U.K. amounting to £2.25 billion worth of new contracts, accounted for 3% of total visible exports. At present the major defence contractors in the U.K. are British firms, but gradually overseas companies are taking some share of the market. For example, Philips, Schlumberger, Hewlett-Packard and Siemens each had contracts with the MOD worth between £5-50m in 1984. Although the MOD has equipment contracts each year with a very large number of companies, a large proportion of the total budget is awarded to a small number. For instance, although the MOD might expect to have as many as 10,000 contractors in one year, in 1984 37% of the equipment budget went to just 12 companies (Breheny, 1986).

We know, then, that defence spending is important to the development of high technology sectors generally and, hence, to individual companies in these sectors; but what do we know of the geography of defence spending, and in particular its role in the Western Crescent? Unfortunately, data on the local distribution of defence equipment expenditure by the MOD is not available. Hence, if we wish to gain some idea of the role of such expenditure in particular local economies we have to deduce whatever we can from a variety of indirect sources. Some information on defence expenditure is available at a regional level in the U.K. The source is Short (1981a) who obtained figures from the MOD for 1977/78. Table 10.21 presents Short's figures for overall defence expenditure and procurement (equipment) expenditure for standard regions. The most obvious feature of these figures is the large share of procurement expenditure taken by the South East (42%) and the South West (15%).

For comparative purposes Table 10.21 also lists overall government expenditure in the regions and the specific provision for regional aid; that is, regional development grants and selective assistance grants given to companies establishing in particular regions. Clearly, procurement expenditure dwarfs regional aid and is generally greatest in those regions not receiving this assistance.

Table 10.22 converts the absolute levels of expenditure in Table 10.21 into a per capita form. Even on this basis, the South West (£95.9) and South

Table 10.22: Defence and Regional Expenditure, Per Capita, by Region, 1977/78, £

Region	Defence Total i	Regional (except Defence) ii	Defence Procurement Contracts iii	Regional Aid Grants iv	Totals i+ii	Totals iii+iv
South East	149.6	833.0	69.8	-	982.6	69.8
South West	258.8	703.5	95.9	1.8	962.3	97.7
East Anglia	133.0	697.2	61.1	-	830.2	61.1
East Midlands	102.1	703.0	56.9	0.4	805.1	57.3
West Midlands	67.3	721.9	36.7	-	789.2	36.7
North West	58.6	824.2	37.2	8.8	882.8	46.0
Yorks/Humber	50.2	771.0	12.7	5.5	821.2	18.2
Northern	77.7	906.5	55.9	45.7	984.2	101.6
Scotland	95.7	1009.9	37.5	21.8	1105.6	59.3
Wales	50.0	896.3	9.2	26.8	946.3	36.0
N Ireland	104.6	1169.5	18.2	32.6	1274.1	50.8

Sources: As for Table 10.21

East (£69.8) gained the highest shares of procurement
and expenditure. This table also combines
procurement expenditure and regional aid on a per
capita basis in order to compare the overall
government support (via contracts and assistance) to
industry in each region. The Northern region fares
best on this basis, with a per capita rate of £101.6,
but the regions ranked next are the South West
(£97.7) and the South East (£69.8). Clearly, these
two relatively prosperous regions gain a major
benefit from government expenditure, and procurement
expenditure in particular.

In recent years (from 1978/9 to 1983/4) defence
spending has increased by 26% in real terms compared
to a rise of only 7% in overall public expenditure
(Treasury, 1984a). Simultaneously there has been a
shift within the defence budget towards purchasing
relatively more equipment and also a shift towards
electronics and aerospace. In 1979 electronics and
aerospace accounted for 17.5% and 28.1% of defence
equipment expenditure; in 1984 these shares had risen
to 20.5% and 30.7% respectively.

Given that a large share of procurement
expenditure is spent in the South East region, and
that an increasing share of the overall budget is
going to the electronics sector, we would expect the
concentration of electronics companies in the
Western Crescent to be taking a substantial slice of
the equipment budget. If this is the case, we might
expect the South East's share of the MOD equipment
budget to have risen since Short's (1981a)
assessment. Unpublished MOD statistics appear to
confirm this. Table 10.23 shows Short's 1977/78
figures alongside equivalent figures for 1983/84
gathered as part of the M4 project. These figures
indicate that the incidence of defence procurement
expenditure is becoming more, not less concentrated,
with the South East's share rising from 41% to 54%.

When the size of the equipment budget (£8,355m
in 1985/6, Ministry of Defence, 1986), the large
share of this accounted for by the South East, the
increasing share being spent in the electronics
sector, and the heavy concentration of electronics
companies in the Western Crescent, the importance of
defence expenditure to firms in the area and to the
Western Crescent phenomenon generally becomes clear.
Not only have the sums of money involved in defence
contracts been large, but they have also been
relatively recession proof, during a period in the
early 1980s when defence spending increased rapidly
while the rest of the economy contracted.

Table 10.23: Defence Procurement Expenditure by Region – Estimates for 1977/78 and 1983/84

Region	Expenditure 1977/78* absolute	per cent	Expenditure 1983/84** absolute	per cent
South East	1,175.6	41.6%	3,747.1	54.0%
South West	410.2	14.5%	763.3	11.0%
East Anglia	111.7	3.9%	208.2	3.0%
East Midlands	213.2	7.5%	346.9	5.0%
West Midlands	189.2	6.7%	277.6	4.0%
North West	242.8	8.6%	763.3	11.0%
Yorkshire/Humberside	61.9	2.2%	138.8	2.0%
Northern	174.2	6.2%	138.8	2.0%
Scotland	195.0	6.9%	416.3	6.0%
Wales	25.6	0.9%	69.4	1.0%
Northern Ireland	27.9	0.9%	69.4	1.0%
Totals	2,827.3	100.0%	6,939.0	100.0%

Sources:
* Short (1981)
** Ministry of Defence. In providing these figures, the Ministry point out that these estimates are based on 85% of contracts and that the missing contracts might affect the regional distribution, possibly depressing the South East share below that shown here. Absolute figures have been calculated from the percentage figures supplied.

The general picture formed by this deduction from indirect sources of the Western Crescent's role in defence contracting is borne out by our survey work. In the survey, half of the firms had defence contracts or subcontracts estimated at over 5% of output (Table 10.24). For seventeen of these firms the figure was over 20%. Not surprisingly the foreign firms tended to have lower shares. The figures from the survey are likely to underestimate the importance of defence contracts to firms as, particularly in smaller companies, the MOD research could be translated into 'spin-offs' for the commercial side of the business (see NEDC, 1983). While there is much interest in technology transfer - or the 'spin-off' from university research to commercial use - defence firms and the MOD receive government funding for R&D over double that of the universities and research councils (Cabinet Office, 1984), although most of this is for development not basic research.

Table 10.24: Defence Sales as Per Cent of Output for High-Technology Plants

		Per Cent Value of Output for Defence Contracts and Subcontracts							
Size of Plant (workers)	Not Signif- icant	5-20		20-50*		50+		Total
100+	2	3	(2)	1	(1)	4	(3)	10
20-99	10	1	(1)	4	(2)	4	(2)	19
1-19	10	1	(0)	3	(1)	1	(0)	15
Total	22	5	(3)	8	(4)	9	(4)	44

Notes: Figures in brackets indicate plants mentioning that the location of DREs was of some significance to them.
* 3 firms answering "high" or "considerable" included here.

We have already suggested that there is a long standing link between electronics companies and the MOD, via the DRE's, in the Western Crescent that goes some way to explaining the current concentration of defence contractors, and hence high technology companies generally, in the area. This historic link is, however, not a sufficient explanation on its own of this persistent and, if anything, increasing concentration. There must be other, complementary reasons for this.
One possible reason for the persistent spatial

Figure 10.4: Ministry of Defence DREs and Contact
Points, 1983.

proximity of defence contractors and DREs, at a time
when in principle such proximity is unnecessary to
business transactions, was uncovered during our
interview survey. This, surprisingly, is the need for
face to face contact between contractors on the one
hand and defence decision-makers, be they MOD
scientists at DREs or MOD administrators, on the
other. Although DREs generally are spread across the
country, the most important ones are concentrated
heavily in the South, and particularly in the South
East. The most important DREs are those with 'local
purchasing power'; that is, they have considerable
autonomy in the making of decisions and spending of
money. Of the eleven DREs with local purchasing power
in 1984 (subsequently rationalised to seven) all
except Malvern (Worcs.) are in the South East or
eastern South West. In addition, as Figure 10.4
demonstrates, MOD 'contact points', to which initial
inquiries concerning defence contracts are
addressed, are also concentrated in similar location
to the major DREs. When asked about the spatial
proximity of their companies to these various defence
decision-makers, our interviewees stressed the
crucial importance of face to face contact in a
business where the use of 'leading edge' technologies
means that constant discussion is essential. Breheny
(1986) has suggested that the spatial proximity of
contractors and defence decision-makers may be a
peculiar outcome of the defence contracting process
in the U.K.

A further reason for bias in defence contracts
towards the south may be the effect of the crossflow
of MOD staff to defence firms. This has reached the
stage whereby a House of Commons Committee expressed
grave concern (Financial Times, 2.12.84). The ex-MOD
staff bring valuable contracts, particularly as most
contracts are based not on competitive tendering, but
on track record and trust that the firm can carry out
the work. It is likely that many of the staff are
resident in the South East (where most of the higher
level MOD jobs are) when they leave the government
service, and are unlikely to wish to move. Hence
firms in the South may have further advantages in
terms of important contacts with potential new staff.

If it is the case that the nature of the defence
contracting process does have distinct spatial
outcomes, it is interesting to speculate on the
effects of recent and future changes to that
contracting process. One such change is in the role
of particular DREs. Since 1984 the number of DREs
designated as centres of excellence in defence

technology has been cut to seven. The naval
establishments have been merged to form the Admiralty
Research Establishment centred at Portsdown near
Portsmouth (at the former Admiralty Surface Weapons
Establishment), although the deputy of the new
establishment will be at Portland near Weymouth (the
former Admiralty Underwater Weapons Establishment).
The cluster of defence firms near these two
establishments is likely to continue, although the
former unit at Teddington, Middlesex is now to be an
outstation (with Dunfermline and Dorset). If
decision making power is moved from Teddington to
Portsdown, it is possible that in the long run firms
could favour the latter location.

A second factor possibly affecting future
developments is the demand by the MOD for more
competition. Only 20% of contracts are awarded after
competitive tender, despite cost savings of 30% found
following the introduction of competition (Defence
Statement, 1984). Although competition would be
limited to keep a strong U.K. base and also to
safeguard long term research and development, it may
make firms more conscious of the high costs of the
southeast, (Breheny and McQuaid, 1984). It is likely
though, that the need for close ties with DREs and
Whitehall would overcome this at least for the R&D
functions. However, if more competition was
accompanied by more contracts going to smaller firms,
then spin-offs to civilian use of military research
could possibly be greater, and more entrepreneurs may
leave the large firms to start up alone.

A further future factor is the impact of major
defence commitments such as Trident and collaboration in the Strategic Defence Initiative ('Star
Wars'). With 45% of the Trident expenditure going to
U.S. firms, the size of this programme may lead to
cuts in other equipment purchases. If this is
accompanied by the intended halt in defence spending
increases, then the huge rises in government defence
equipment procurement that have been benefiting the
South of England, and the Western Crescent in
particular, may slow down appreciably; although the
absolute levels will remain far greater than anywhere
else in the country. There may also be two, possibly
counteracting, consequences. If defence budgets
tighten the large firms may subcontract less of their
contracts, possibly harming smaller companies.
However, such a cut back (or halt in expansion) may
force the companies to increase spin-offs to civilian
uses, as well as attempting to increase arms exports.

In summary, defence equipment procurement has

had a strong bias towards the South East and South West, far in excess of regional aid given to firms in other areas. In recent years particularly, such expenditure has greatly increased, while aid to firms in other regions has declined. The initial siting of defence plants and their subsequent expansion can be partly traced to government policies of locating DREs and major procurement offices in the South of England. This link has been sustained, for a number of reasons, to the great benefit of defence-related high technology companies in the Western Crescent.

Conclusion

The first half of the paper described the geography of high technology industries in Great Britain, and set it within the context of national and regional employment change. Although each individual high technology industry exhibited a unique geography of employment change, the counties around London's 'Western Crescent' (particularly, Hertfordshire, Berkshire and Hampshire) fared consistently well in most of the industries. The areas further west along the 'M4 Corridor' showed no major employment gains, and in fact many areas lost considerable numbers of high technology industry jobs. The second half of the paper discussed the key factors apparently determining the genesis of high technology industries in the Western Crescent. The major factors appear to be: the early development of the electrical industry in west London, superior communications, skilled labour availability, the 'spawning' of new firms, and defence spending.

As a preliminary attempt at explaining the reasons for the growth of the Western Crescent as a concentration of high tech activity this paper has adopted this simple 'factors of location' approach. Other, more theoretically-based approaches have been adopted in assessing the growth and location of high technology industries, but these have usually discussed high tech in abstract rather than any particular local concentration. As Sayer (1985) argues, there will always be strong local contingencies that will confound attempts to apply particular theoretical approaches. Thus, such explanations will always require thorough empirical work.

There is no doubt that such approaches – traditional cost-minimisation theory, product cycle theory, spatial divisions of labour, for example – do help in understanding the Western Crescent

Table 10.25: The layers of 'cumulative advantage' in the Western Crescent

DATE	1920	1930	1940	1950	1960	1970	1980
EVENT	Development of electrical industry in W. London		→ Conversion to military production	→ Decentralisation to south and south-west London		→ Spawning of small firms westwards	→ Continued growth of small firms
		Growth of GRE's west and south west of London		→ 'Cold war' weapons development		→ Growth of electronic weaponry	→ Increase in defence equipment budget
				Opening of Heathrow			
			Abercrombie's Greater London Plan	→ Designation of Bracknell New Town		→ Strategic Plan for the South East	→ Growth area status
						Opening of M4, M3, M40 and 125 rail system	→ Opening of M25
						Influx of U.S. companies	→ Growth of U.S. companies
EFFECT	Original high tech concentration		Establishment of local military-ind. complex	Start of planned growth		Major infra-structure boost	'Western Crescent' recognised
						Develops as internationally important location	

phenomenon. It is clear, for instance, from earlier discussion that the area functions as a 'control' location within the UK and Europe, and that this role is if anything becoming more pronounced. Certainly, this theoretical perspective helps us to understand the current function of the area and the nature of its recent history. It does not help, however, in explaining the crucial early development of the area; overseas multinationals moved to take advantage of the concentration, not to create it.

How, then, do we piece together the various strands - the early origins, infrastructure investments, planning policy, spawning, defence expenditure, etc - of an explanation of the growth of the 'Western Crescent'? Possibly a useful way forward might be to regard these as 'layers of advantage' that have developed, some purposefully, others accidentally, since the beginning of the century. These layers, illustrated in Table 10.25, have tended to have a strong reinforcing effect, giving a cumulative and powerful advantage to the area, along the lines of Myrdal's (1957) notion of 'cumulative causation'. Just as changes in many depressed parts of the UK have had cumulative and reinforcing deleterious effects, so areas like Berkshire have experienced the opposite beneficial effect.

This 'cumulative advantage' view is useful because it reflects the importance of understanding detailed historical developments and of the locally contingent nature of development; no single theory is likely to explain such a complex detailed history. The view is also useful because it immediately focuses our attention on the degree to which these different cumulative histories affect the economic prospects of different localities. The depth of the advantage accumulated by areas like Berkshire over a long period suggest that they will inevitably continue to maintain that advantage. With a strong but diverse representation in new industrial sectors, with a large pool of skilled labour, with an innovative climate, with excellent infrastructure provision, and so on, the area is likely to prosper relatively for some time to come.

Conversely, the depth of disadvantage now accumulated by some depressed areas of the UK is such that they are unlikely to see any short term reversal in their industrial fortunes. It is this depth of disadvantage that militates against the success of, what are in truth, often superficial local economic initiatives.

Ultimately, it may be in this field - the

development of measures to reverse the fortunes of depressed localities - that this and other studies of the genesis of particular concentrations of high technology industry can make a practical contribution. In many cases the contribution might be a rather depressing one, by pointing out the difficulty of replicating layers of 'cumulative advantage' in the short term. In a small number of cases, some such layers may already exist, but not have been recognised or taken advantage of; here - bearing in mind the large role played, albeit often unwittingly by the public sector in the development of the Western Crescent - replication may not be so difficult.

References

Abercrombie, P., (1944), Greater London Plan 1944, London: HMSO

Ball, N. and M. Leitenberg (eds), (1983), The Structure of the Defence Industry, London: Croom Helm

Barnaby, F. (1982) Microelectronics in War, in Fredrichs, G. and Schaff, A. (eds) Microelectronics and Society, A Report of the Club of Rome, 243-72

Berkshire County Council, (1985), Survey of Employers, Reading, Berkshire C.C.

Braun, E. and P. Senker, (1982), New Technology & Employment, London: Manpower Services Commission

Breheny, M., (1986), 'Contacts and Contracts: Defence Procurement and Local Economic Development in Britain', paper presented to the Anglo-American Workshop on the Growth and Development of High Technology Industry, Churchill College, Cambridge, June

Breheny, M.J. and McQuaid, R.W., (1984) The Genesis of High Technology and Industry in the M4 Corridor: A Preliminary View, paper given at the Regional Science Association Annual Conference University of Kent at Canterbury, 5-7th September 1984

Brown, C.M., (1966), The Industry of the London Region, in Martin, J.E., Greater London: An Industrial Geography, London: G. Bell & Sons

Cabinet Office, (1984), Annual Review of Government Funded R&D 1984, London: HMSO

Castells, M. (ed.), (1985), High Technology, Space and Society, Beverly Hills, Sage

Economist, (1984), Property's new high ground, The Economist, November 3 1984, 74

Ellin, D. and Gillespie, A. (1983), Production Industries and Job Creation in Great Britain, Working Note, C.U.R.D.S., University of Newcastle-upon-Tyne.

Engineering Industry Training Board, (1984), Does
the M4 Corridor Exist? Research note, E.I.T.B.,
London
Gibbs, D. (1983), The Spatial Incidence of High
Technology Industry and Technological Change,
Working Note, C.U.R.D.S., University of Newcastle-
upon-Tyne
Glasmeier, A., (1985), Innovative Manufacturing
Industry: Spatial Incidence in the United States, in
Castells, M. (ed.) High Technology, Space and
Society, Beverly Hills, Sage, 55-80
Hall, P., Thomas, R., Gracey, H. and R. Drewett,
(1973), The Containment of Urban England, London:
George Allen & Unwin
Hall, P. and Markusen, A. (1982), Innovation and
Regional Growth: A Proposal for Research, Institute
of Urban and Regional Development, University of
California, Berkeley, unpublished
Hall, P. Breheny, M., McQuaid, R. and Hart, D. (1987
forthcoming), Western Sunrise: The Genesis and
Growth of Britain's Major High Tech Corridor, London,
George Allen and Unwin
Hampshire County Council, (1984), High Technology
Industry in Hampshire, Strategic Planning Paper 15,
Winchester, Hampshire C.C
Howells, J. (1984), The Location of Research and
Development: Some observations and evidence from
Britain, Regional Studies, 18, 1, 13-30
Hymer, S.H., (1975), The Multinational Corporation
and the Law of Uneven Development, in Radice, H.
(ed.), International Firms and Modern Imperialism,
Harmondsworth, Penguin, 37-62
IPCS (Institute of Professional Civil Servants),
(1984) Nuclear Arms, Defence Spending and Jobs,
Mimeo, London, IPCS.
Joint Team, (1971), Strategic Plan for the South
East, London, HMSO.
Kaldor, Mary, (1982), The Baroque Arsenal, London:
Deutsch
Keeble, D., (1968), Industrial Decentralisation and
the Metropolis: the North-West London Case,
Transactions of the Institute of British Geographers
44, 1-54
Keeble, D., (1976), Industrial location and planning
in the U.K., London: Methuen
Lavington, S., (1980), Early British Computers,
Manchester: Manchester University Press.
Law, C.M. (1983), The Defence Sector in British
Regional Development, Geoforum 14, 169-84
Lovering, J. (1983), Regional Intervention, Defence
Industries and the Structuring of Space in Britain,

the case of Bristol and South Wales, paper given to the Cardiff Branch, Regional Studies Association, October 1983

Martin, J.E., (1966), Greater London: An Industrial Geography, London, G. Bell & Sons

McQuaid, R.W. and Langridge, R. (1984), Defining High Technology Industry, paper presented to the annual conference of the British Section of the Regional Science Association, Canterbury, Kent, September.

Ministry of Defence, (1986), Statement on the Defence Estimates, Cmnd 9763, London, HMSO.

Ministry of Defence, undated, Selling to the MOD, London: Ministry of Defence

Myrdal, G., (1957), Economic Theory and Underdeveloped Regions, London, Duckworth

NEDC (National Economic Development Council), (1984), Civil Exploitation of Defence Technology, London: NEDC

Plessey, (1983), Report & Accounts 1983, London: Plessey Co.

Sayer, A., (1985), Industry and Space; A Sympathetic Critique of Radical Research, Environment and Planning D: Society and Space, 3, 3-29

Saxenian, A., (1983), The Urban Contradictions of Silicon Valley, International Journal of Urban and Regional Science, 7, 237-61

Segal Quince & Partners, (1985), The Cambridge Phenomenon: The Growth of High Technology Industry in a University Town, Cambridge: Segal Quince & Partners

SERPLAN (South East Regional Planning Conference), (1983), Employment Monitor for 1982/3, SCI812, London: SERPLAN

SERPLAN (South East Regional Planning Council), (1984), Employment Change in the South East Region: 1971-1984, London, SERPLAN

Short, J., (1981a) Defence Spending in the U.K. Regions, Regional Studies 15, 101-10

Short, J., (1981b) Public Expenditure and Taxation in the UK Regions, London: Gower

Street, S. and Beasley, J., Investing in Defence, London, de Zoete Bevan

Treasury, (1984a), The Next Ten Years: Public Expenditure and Taxation into the 1990s, Cmnd 9189, London: HM Treasury

Treasury (1984b) Public Expenditure Plans 1985-6 to 1987-88, Cmnd 9428, London: HMSO.

INDEX

aeronautics
French industry 216-18
American semiconductor industry 37-79
availability of skilled labour
force 56-7
California, emergence in 49-
51
capital deepening 58-69
concentrated territorial development, predicaments of 58
continued managerial and
technical domination at core
72-3
core complex, formation of
49
development of Silicon Valley
production complex 51-3
development of structure 44
domestic semiconductor shipments by major product class
52
emergence of socially and
spatially segregated labour
markets 47-8
emergent cores within regional sub-divisions 74
geographical origin of semiconductor shipments 46
growth of 37-79
horizontal and vertical
breakdown of production
activities 53
internal necessary relations
43
internationalisation 37-9

determinants of 58-9, 59-
61
Scotland 57-71, see also
Scotland
South-East Asia 62-7, see
also South-East Asia
labour processes 45-49
location process, determinants of 60
markets 58-9
new spatial political economy
71-4
occupational structure
Table 48
polarisation in skill structure
of labour force 47
reorganisation of labour
process, and 48-9
Santa Clara High Technology
employment 55
Shockley, William 50, 51
Silicon Valley phenomenon 49
social reproduction of labour
force 53-7
spatial division of labour
Scotland 67-71, see also
Scotland
South-East Asia 62-7, see
also South-East Asia
spatial reproduction of labour
force 53-7
spatial theory 41-2
technical change 45-9
effect of 73-4
trade union organisation, and
56

355